Table of Contents

A GLOBAL VILLAGE BOY

By
Emmanuel Saffa Abdulai

DEDICATION

This novel is dedicated to my cherished family: Sadiya, my beloved, and our two children, Daniel Musa Abdulai and Nyaliima Rebecca Abdulai. I also honor the memory of my dear friend, Arthur E. E. Smith, a lecturer at Fourah Bay College, University of Sierra Leone, whose passion for literature and teachings remain cherished.

CHAPTER 1

THE BIRTH OF A VILLAGE BOY

As the sun peeked through the window, Hindolo stirred from his deep slumber. He rubbed his eyes as he sat up and shielded his face from the sun's unrelenting rays pouring through the window. Though he usually loved early mornings, he felt exhausted and sluggish today.

But wait, he realized suddenly, today was his seventh birthday! A rush of delight rushed through him as he remembered the surprise his father, Kamoh, had sprung on him at midnight. A heavenly scent wafted into his nostrils as he sat in his room, engrossed in a book and his father had walked in with a scrumptious ginger cake with seven candles lighting it. Hindolo had been beside himself with joy. He'd blown out the candles, and afterward, they had stayed up late into the night, devouring the delicacy while reminiscing about his mother and exchanging hearty laughs.

Hindolo couldn't help the grin that spread across his face. How could he not be happy? It was his birthday today!

Suddenly a pebble flew in through the bars in his window onto his bed. Hindolo looked at it bewildered. Then another flew in, and then another, each pebble followed by giggles and snorts. Hindolo grinned. This was his friend's way of waking him up and forcing him to come out.

He picked up the pebbles and tossed them out.

"I'm coming," he called out.

It was his birthday today, and that meant an entire day of being the star of the village. He excitedly put on his clothes and ran outside, where he was greeted by his friends, Amara and Hawa.

"Happy birthday!" they both cried out.

"Thanks!" said Hindolo as he grinned back.

It was early morning. The sun had risen only a few minutes ago, bathing their village in a golden haze. The sky was pale blue streaked with tinges of pink and gold here and there, as well as white cotton clouds. Hindolo's house was on the left, towards the backside of the village. Amara lived two huts down, while Hawa lived a bit further ahead.

In the days of old, the village of Luawa had been home to circular huts, much like most other African villages. However, as the world began to change, so too did the people of Luawa. It was Hawa's grandfather who had led the charge for change.

He had been sent away to study at a prestigious university and had returned as an accomplished engineer, ready to change the way they lived. With his help, the village made modern houses that were fully equipped with electricity and a water system, a frill unheard of in their time.

Hawa's grandfather had been later chosen to take the place of a village elder, a respectable and honorable position that placed his house at the heart of Luawa. However, it was more than just a title; it was a duty, a sacred responsibility to ensure the safety and wellbeing of the people that resided in the village. And for Hawa's grandfather, it was a responsibility he held close to his heart as he tirelessly worked himself to the bone to make Luawa a haven for all those who called it home.

He was a fair and just man, and everyone loved him.

Despite the initiation of modern housing, the village of Luawa remained committed to preserving its rich traditions and customs. They refused to change much else. The village square continued to serve as a gathering place for the elders to assemble and deliberate on important matters.

Luawa was a very close-knit society, with the elders retaining the ultimate power of decision-making. They were highly protective of their history, eager to pass on their principles and beliefs to the next generation, and adamant that their ideals were not lost in the process of modernization.

Despite the winds of change pouring through their village, Luawa clung to its original and one-of-a-kind layout, which was created to mimic the human body. The elder's home stood at the heart of the village, an expression of their loyalty to tradition and unchanging commitment to the well-being of their people.

Their houses were built of clay and wood with grass-thatched roofs. The farms had been flourishing since Hawa's grandfather built an irrigation system. The village was green and lush. Crops were growing faster than the speed of light, and the people were prospering.

The people of Luawa had thrived so much under Hawa's grandfather that they had built their very own medical center for the sick, which was a far cry from the days when they would have to travel miles to towns for medical attention.

Hindolo ran through the village with his friends in tow. By now, people were fully awake and heading to their farms or the village square to chat with their neighbors and friends before going to work. The village square was in the center of Luawa. It was a common space shared by each inhabitant of their village. It was buzzing with the chatter of people here and there, bargaining in shops, sitting with their friends.

It was a Saturday, which meant most people were relaxing today. As they walked through the village square, people everywhere started wishing Hindolo a happy birthday.

Hindolo couldn't stop beaming.

"Hey, Hindolo. Come here!" yelled the baker from his shop.

Hindolo and his friends ran up to him excitedly. The baker loved kids, and he gave free dessert on birthdays.

"Happy birthday, Ndopui. Here is some cake for all three of you. Have fun!" he smiled.

"Thank you!" cried out all three boys.

The baker laughed, as did the people at his shop, each wishing Hindolo a happy birthday.

"Hey, look, there's Sattu!" said Amara.

Hindolo turned around so quickly that his neck cricked. There she was - Sattu, the girl he had liked since kindergarten. When Hawa and Amara noticed how he had responded, they burst out laughing, brutally taunting him for his crush.

Hindolo was a meek and timid boy. He stood tall for his age and had thick, wavy hair. But he was so self-conscious about it that he shaved his head frequently. Despite his self-doubt, his deep, dark eyes and lush lips were eye-catching features that often drew attention. His father used to say he looked exactly like his mother, and Hindolo treasured that compliment. Although he had never met his mother, the knowledge that he carried a part of her within him and bore a resemblance to her was a fact that Hindolo held close to his heart.

Sattu, on the other hand, was a beautiful girl. Her eyes were green like the sea, and her chocolate brown curly hair fell past her shoulders all the way down to her hips. They would reflect against the sun and bring out her sandy skin color. As soon as she spotted him, she waved at Hindolo and ran towards them while Hindolo stood there frozen, too nervous to say anything.

"Happy birthday, Hindolo!" she squealed.

"Thank you," replied a nervous Hindolo.

"We're just walking around having cake. Do you want to join us?" asked Amara hopefully.

Sattu shook her head and replied, "No, I have to go to the upatam with my mama."

"Okay, see you later then." Shrugged Hawa as she smiled at them before running back to the upatam where her mama was.

The upatam was a sacred space dedicated just to the women of the village, and no man was permitted to set foot there.

Hindolo had wanted Sattu to stay, but even if she couldn't, he didn't mind; at least she had remembered his birthday and wished him.

Hindolo and his friends raced through their village's twisting trails, racing against one another, their joyous laughter and whoops of glee booming through the village like a choir of wild monsters. Running and falling, they finally arrived at their destination: the river.

The river, which flowed softly beside their village, had been a lifeline for its residents. The people depended on this water for everything. However, beyond its glistening waters was a large forest filled with all sorts of animals, both big and small. The people of the village had shared these waters with the animals for as long as anybody could remember, with each group respecting the other's borders and staying on their own side of the river.

The forest was a realm that excited the young ones more than anything else. For them, it was a realm of mystery and magic. It was home to herds of grand elephants and swift monkeys who leaped from branch to branch with ease and often played tricks on the villagers. Even the huge cats of the forest, the lions, and the cheetahs supposedly inhabited its depths, but they had rarely ever been spotted.

Despite their scary reputation, the woodland creatures seldom presented as a danger to the residents. An inquisitive new born calf who did not understand the rules of their land might occasionally come into the settlement, sparking a flurry of excitement and astonishment among its people. However, the animal was always safely escorted back to its true home deep in the untamed forest.

The Society Bush, a hallowed spot wrapped in mystery and tradition, was located deep in the belly of the forest. Throughout the years, it had remained a place of unparalleled importance to the inhabitants of the village. It was a rite of passage that marked the transition from boyhood to manhood.

Only the most daring young boys were brave enough to enter the Society Bush, which was supposed to be a site of immense might and was said to be enchanted. The boys would be taught the competencies and knowledge required to become a true man here, under the vigilant eye of the elders supervising them.

The teenage boys would be prepared for many different things during their training. After all, a kid must be taught many things before he can be deemed fit to be a man and care for his family. The boys were taught essential skills like hunting, trailing prey for food, and how to shield themselves against the perils of the forest. They were taught how to converse with authority and confidence, how to have command over a room with merely your tone, and how, in times of trouble, they could lead their people with both knowledge and wisdom.

They were taught to be courteous, the value of responsibility, the importance of hard work, and the benefits of leading a life of integrity and respect.

Even the most fundamental skills were taught with tremendous attention to detail in the Society Bush. The boys would learn to wash themselves using only a water bottle, a simple yet necessary skill to have when you live in a village.

A series of marks are put on the back of the graduate to demonstrate that he has graduated from this secret society. Graduation was considered a monumental event and the entire village would celebrate. Every family anticipates their children will join the poro society. Kamoh also wanted Hindolo to join not just for the sake of it but also to transition into a man.

Young girls take pride in talking to a graduate from the Poro society, and a new self-esteems and society pride is bestowed on the graduate.

The Poro society is not for the faint-hearted. It's a grueling 90-day initiation that involves everything from rigorous drills to hunting lessons and lectures. In fact, some initiates have even lost their lives during the process, a fact that's a source of major concern for the parents of the boys who dare enter the society.

To prepare for their sons' initiation, families are required to save and store rice, yam, fish, and bush meat, among other things. They have to cook for the entire batch of initiates and their mentors and lecturers, which is no small feat. The cost of the initiation is high, but the pride and prestige that came with being a member of the poro society made it all worth it.

At the end of initiation, a flamboyant ceremony is lavished. New dresses were prepared, shoes bought, and chairs were made to display the successful ones.

Hindolo and his friends sat on the riverbanks, transfixed by the scene in front of them, quietly watching. The elephant herd emerged from the forest, one after the other, striding with such surefootedness it seemed as if the river was their kingdom and they were the rulers. The newborn elephants were the highlight of the scene as they jumped around in the water and frolicked here and there, full of enthusiasm and wonder.

Hindolo was particularly taken with a newborn elephant. It appeared to be striving to show off by swimming across the river to the side where Hindolo and his friends were seated. Hindolo couldn't help but smile at it. The newborn elephant's joy was infectious.

The happy moment was cut short as the mother elephant let out a thunderous horn. The young elephant came to a halt and returned to his mother's side, where she gently cleaned him with her trunk and rooted away as if chastising him for his careless behavior.

As Hindolo sat quietly on the riverbank, he couldn't help but feel a pang of sadness in his heart. Watching the baby elephant being bathed by his mother only reminded him of the mother he had lost. His own mother had passed away on his birthday, and it always made the celebration bittersweet. While his friends chattered away, he couldn't help but feel a sense of longing for the one person who would understand his feelings.

Meanwhile, Hindolo's father, Kamoh was also feeling the weight of the loss. He missed his wife dearly and had been visiting her grave on Hindolo's birthday every year since she passed away. He couldn't bear the thought of not honoring her memory on this special day.

They had been together for nearly ten years before God blessed them with the child they had so desperately desired. However, Hindolo knew in his heart that if Kamoh knew having a child meant losing his beloved wife, he would have never prayed for one in the first place.

Hindolo's mother, Sattuma, died the day after he was born. She had fought long and hard throughout labor to deliver Hindolo safely, but she had succumbed to her battle against the injuries sustained during childbirth.

Her pregnancy had been difficult, and she had feared she wouldn't make it, yet such is the love of a mother. All she cared about was the life within her womb, not the threats it posed to her life.

Sattuma bled profusely, and the blood loss was just too much for anybody to bear. However, neither the pain nor the prospect of dying seemed to faze her. Sattuma had merely smiled and murmured, "Even if I only get one hour with my baby and my husband, I will die a lucky girl tonight," when the midwife warned her and Kamoh that she would not survive the night.

Kamoh, however, was beside himself. Men in their tribe did not express such sentiments, but Kamoh was undeterred. He had always stood out from the crowd and was different from most men. Kamoh was educated, and he understood the value of showing love and affection.

While most men avoided showing their emotional side since men who showed emotion were considered weak, Kamoh had spent the previous ten years squandering all the love that he had for his wife and then his child. He couldn't spend a day without seeing her, and they hadn't just been married for ten years; they had also been friends since childhood since they grew up in the same village.

Sattuma only smiled as Kamoh sobbed. She laid on her back with Hindolo pressed against her bare chest while Kamoh sat on the floor next to his wife.

Hindolo had heard this story a million times from his father. His mother embraced him and whispered to Kamoh, "Take care of our kipusa mchanga. He's destined for greatness. You have to be strong for him. You have to be both his mama and papa. He will be our legacy. He will carry our names on when the rest of the world forgets."

She cupped Kamoh's face, and Kamoh held her hands against his as he watched the light leave her eyes as the angels came and took her over to the other side. Death is a massive tragedy, but when someone you love more than life itself passes, the pain is unimaginable. Kamoh had never felt such agony in his life. Not even the rigors of his Poro society's demands.

Since then, Kamoh had slowly let himself go. He couldn't move past her death, and he stopped taking care of himself. Other than his father and aunt Demy, Hindolo had no other family.

His aunt Demy was a wonderful woman who had left the village long ago. She now lived in the city and would visit them a few times a year whenever she had the time. While Kamoh did his best to be a good parent, he was mostly lost in his own thoughts.

Sometimes, when he couldn't sleep at night, Hindolo would sneak out of his room, and he would always find Kamoh sitting in his chair, drowning his sorrows in alcohol, staring at a picture of his wife.

He often wondered if his father perhaps blamed him for his mother's death. He always showered Hindolo with a lot of love, more love than anyone could imagine, which is why Hindolo wished he could do something for him to make it easier.

His aunt always told him to leave his father alone. She would say, "Time heals all wounds, and what wounds time doesn't heal, we have to heal them on our own. You can't help your father if he does not wish to be helped."

"It's almost lunchtime; let's go back to the village now." Said Hawa, all of a sudden, breaking Hindolo's chain of thought.

Hindolo nodded and replied, "Let's go."

They walked towards the village and went to their homes to eat. Hindolo was excited to eat dinner with his dad. Kamoh, however, was nowhere to be found.

Hindolo sighed to himself. Kamoh was probably still at the graveyard, so he decided to make something for himself. There were some leftovers from yesterday, which Hindolo decided to eat now. He finished eating and went to take a nap, tired from the early morning adventures he had.

Hindolo's eyes slowly opened, still heavy with sleep. But then someone shook him so hard that he felt like he had been jolted out of a dream. It was his father's friend, and from the urgency in his voice, Hindolo knew that something was wrong.

"Hindolo, wake up!"

"Come quick."

He quickly got up and followed the man, his heart racing with anxiety.

As they ran through the village, Hindolo's mind raced with possibilities of what could have happened. Had there been an accident? Was someone sick? When they arrived at the medical center, his father's friend turned to him with a solemn expression that made Hindolo's heart drop.

"What's happening?" asked a scared Hindolo.

"I think I better let the elder talk to you." He replied nervously.

Just then, the elder came out.

"My dear boy, we have some terrible news for you. Let's sit down." He said as he motioned towards a bench.

Hindolo sat down, his heart beating fast.

"Your father is no more. We found him in the cemetery, collapsed over your mother's grave. He'd been unwell for a long time but hadn't informed anyone. Even though the doctors here had cautioned him, they had no idea how far his disease had advanced. It eventually got the best of him today, and he passed away from his disease. We spoke with your aunt. Your father chose her as your caretaker in the event that something happened to him, and she is on her way to pick you up. You can say your final farewells right now. The funeral will be held by sunset." Kabba, one of the village elders, explained.

And that was that.

Hindolo was an orphan now. On the same birthday, he had now lost both his parents. He couldn't stop the tears from flowing as the elder took him to see his father. He looked the same. His lips were bluish, and his eyelids were closed. The tears streamed down his face, and he couldn't hold them back. His father looked peaceful, almost as if he was sleeping, but Hindolo knew the truth. He was an orphan now, left to navigate life without his parents by his side. The words he wanted to say were trapped in his throat, and he felt like he couldn't breathe. Finally, he leaned in and pressed a gentle kiss on his father's cheek, whispering his final goodbye. As he turned to leave, he couldn't help but feel like his world had come crashing down around him.

The funeral procession was a somber affair, with the villagers dressed in black and the sound of mournful music filling the air.

16

The beating of drums echoed in the distance as the procession made their way to the burial site.

As they arrived, they began to perform the traditional ritual of mende. The elders chanted prayers to Leve, the god of the tribesmen in Sierra Leone. His name was shrouded in mystery, with many believing that it was a far older name than they could comprehend.

'This is what Leve passed down to us long ago,' the Mende says of customary practice.

As Ngewo, the sky god, is detached from human matters, they think that natural events are manifestations of the deity's might. As a result, he sends rain to his 'wife,' the earth. The spirits—ancestral spirits and genii, dyinyinga—stand between Ngewo and humans.

The latter are connected with rivers, woods, and rocks, whilst the former is associated with rituals aiming to allow connection between men and the sky deity.

After the ceremonies of tindyamei, or "crossing the water," the departed spirit enters the land of the dead. The deceased is aided on his trip by the artifacts put in the tomb, which is referred to as a "home" because the spirits "on the other side" expect to receive gifts from the visitor. Denying a person's funeral rituals is equivalent to forcing his spirit to remain on earth and, as a result, to be plagued by it.

Despite being part of the village community, Hindolo never fully appreciated the significance of his culture's funeral rites. The elders tried to explain it to him, but he couldn't bring himself to focus, immersed in his own thoughts and pain. All he wanted was for it to be over, for the anguish to go away, and for everything to go back to normal. But he knew deep down that this would never happen.

Hindolo was sitting outside on the porch of the elder's house, staring blankly into space, his mind numb with the events of the past few hours. His father was gone, and he was alone in the world. He didn't even know what he was feeling or how to express it.

As he sat there, lost in thought, the villagers came by to offer their condolences. Some brought food, others brought gifts, but it all seemed like a blur to Hindolo. He couldn't bring himself to say thank you, let alone engage in any meaningful conversation.

The next morning, Hindolo's aunt arrived. She had missed the funeral, and Hindolo wondered if it mattered at all. She went to visit her brother's grave, and when she returned, she hugged Hindolo tightly. It was a long, warm embrace that made Hindolo feel like he wasn't alone.

His aunt told him that she would take care of him now and that they would start a new life together in the city. It was a scary idea to be away from the village he had never left, but it was also a chance for a fresh start, away from the memories of the past. He had no one.

She let him say his goodbyes before he packed his things and left to start a new life in the city.

The city was a world away from the village where Hindolo grew up. It was a bustling, sprawling metropolis with towering skyscrapers that seemed to stretch up to the heavens. The streets were crowded with people, cars, and bicycles, all racing to get to their destinations on time. Everywhere Hindolo looked, there was movement, noise, and activity. It was overwhelming.

Living in the city was a stark contrast to the slow and simple life Hindolo had known in the village. Here, everything moved at lightning speed, and trust was a rare commodity. Hindolo quickly learned that he had to be on his guard at all times, especially when he was alone. He watched his aunt navigate the city streets with ease, avoiding the pickpockets and scammers who lurked in the shadows.

During the day, Hindolo was often alone in their small apartment while his aunt worked. But he soon discovered that he was not the only child in the building left to fend for himself. The community was full of kids his age, all in the same boat, with working parents who had no choice but to leave them home alone for hours on end.

Hindolo had to adjust to a lot of different things. The loud noises of the city, the horrible nature of some people, the dirt, the horrible air, the indifference. He had never seen anything like it, but he quickly learned that in order to thrive, you need to survive.

The city was filled with many different laws and regulations, but one rule that Hindolo absolutely despised was the ban on playing football on the streets. It felt like an infringement on his freedom, an attempt to stifle his joy and energy. But Hindolo was not one to sit back and accept the status quo. He was determined to fight against this rule in whatever way he could, even if it meant breaking the law. To hell with this ridiculous law.

His friends were equally as rebellious friends as him, and they would often take to the streets with a football, kicking the ball back and forth with laughter and shouts of joy. They knew the risks of being caught. They ran, dodged, and scored, all while keeping an eye out for any police officers who might try to arrest them.

Police officers and other security men were often chasing them, but they knew their neighborhood quite well and managed to get away each time, pulling a quick one over the cops.

They would usually go and play 'war' in the adjoining woods and bush region, imitating movies like Rambo and Commando in the hillside hamlet with schools nearby.

Their war games were both fascinating and brutal. The boys would become so engrossed in the activities that they would forget it was just a game. It was their own personal take on Mortal Kombat.

On one particular day, Hindolo was busy painting his face and hiding behind an old tree in the forest when his friend Diallo appeared from behind him and shouted, "Surprise attack!"

He used his entire body weight to push Hindolo onto the ground.

"Ow, what was that for?"

"Eh, I just felt like it," he replied.

Hindolo gave him an unamused look. He rolled his eyes.

"When are the others coming?" He inquired.

"Don't know. They should be here soon though".

Hindolo nodded and then went back to painting his face.

Their friends soon appeared, and they immediately got into an intensive combat battle unlike any other. Their ecstatic yelling noises were audible across the forest and could be heard everywhere. One would think they were in a real battle of some sort. They were making angry and exaggerated noises while collapsing on the ground and simulating gun sounds to make it as realiztic as possible.

'Bang!' 'Bang!' 'Bang!' 'Bang

Hindolo's team was losing terribly, and they couldn't figure out how to deal with the opposing team, who was playing exceptionally well, and their strategy, which was throwing mud balls at them was working.

While fleeing, his squad ducked and hid themselves to avoid being attacked, thinking they were in the clear, when the other team's opponent pounced on them and flung a massive mud ball straight at them.

"Hooray! Kagiso was victorious in the battle. He infiltrated the team and tossed the mud ball, winning us the match!"

Hindolo sighed as he cleaned the muck off his clothing. He had suffered a rather devastating and humiliating defeat.

This was normally how their days went; teams of two would be formed, and the losing team would have to "serve" the winning team by shining their shoes and lugging their belongings everywhere. They were basically the winning team's servants for the day. This had become a little custom for them, and they loved it.

Hindolo and his friends were like a pack of wild animals, always looking for new adventures. They relished the thrill of danger, and their deviant behavior had no limitations. Their pranks often landed them in hot water, and they were no strangers to the long arm of the law. They were skilled at eluding cops, dashing through the city's small alleys and vanishing into the shadows like wraiths.

They would also steal small things and scheme to take fresh-grown mangos and oranges from the neighborhood and numerous compounds. They would break into houses or stores and take what they pleased. While one friend would be distracting the owners, the others would swipe what they had. They did this not only to local sellers, but they also stole from individual residences and were quite rebellious and constantly up to silly pranks.

They would often get into fights with other kids as well, and Hindolo, being smaller than the other kids, would often end up getting beaten.

The more time passed, the more rebellious Hindolo became. He and his friends went from breaking the rules by playing football to breaking into people's houses or stores and stealing small things. No one could prove it was them, but everyone knew it. He and his friends were regarded as outcasts, and no parent wanted their child to become friends with them.

As one could expect, the local cops were inept. The state's circumstances were also deplorable, which explained why the police were so bad in such a failing state. There was a one-party rule, and the police were underpaid; they were frequently seen raiding and detaining children in order to extract money from their parents. To make matters worse, the guilty party was imprisoned indefinitely if they did not pay.

His aunt was usually too tired or too preoccupied with work to properly check on him, and her husband didn't care. He despised having another mouth to feed since it was money out of his own wallet. Hindolo mostly stayed out of the house, walking about and stealing food so his aunt and uncle wouldn't have to feed him. He was afraid they might kick him out or that his aunt would get tired of him.

Hindolo would lay on his improvised bed of old rags, staring up at the stars that looked so far away as the moonlight flooded through his window. He yearned for his village, his friends, and, most of all, his father during those times. The recollections of his past would come back to him like a roaring river during these lonely hours. He, however, knew that he had to accept his new situation. The city had become his home, and he had to learn to live there, no matter how difficult it was.

CHAPTER 2

THE CRADLE OF VILLAGE BOYS CIVILIZATION

Hindolo walked along the loud streets. He had been out since morning and hadn't eaten anything. He had school in the morning, and he had been quite excited about it.

It was the one thing he had been looking forward to since he came to the city: education.

His father prioritized his education a lot. He always reminded Hindolo of the importance of being well-educated and how important it was in order to succeed in the world, but his father would also tell him constantly how success came in different forms.

His father would say, "Hindolo, do you think I'm rich and successful?"

Hindolo looked at him uneasily, unsure of how to answer that question.

"Don't be afraid; there's no right or wrong answer. Tell me what you genuinely believe," his father assured him.

"I don't think we're rich and successful, papa. Aren't rich and successful those people who have big cars and big houses and wear expensive clothes?" asked Hindolo meekly.

"So you think richness and success stem from how much money you have?" asked his father in turn.

"I mean, isn't that what it means?" replied Hindolo, confused.

His father shook his head and laughed, "No, my son. These are ideas filled in our heads by those who do not know any better. We are all rich and successful in our own ways, and we each have different measures of success.

For instance, I do not have big cars or a huge house, but I am rich and successful in my own way. I was successful in raising a family and having a beautiful wife who loved me my whole life. She gave me a beautiful son and memories that I will cherish for the rest of my life. Those memories are my riches.

You see, Hindolo, I do not need money to be rich or successful. Monetarily, I have enough to take care of you and save enough to send you to a good university. Yes, I could squander all that to build a bigger house or to buy a bigger car, but those would not be my riches. I have no need for such things. As long as I have you and your mom's memories, I have everything.

Do you think T'challa is successful?"

Hindolo nodded and replied, "He's the richest man in our village, so of course."

Hindolo replied, "What if I told you he does not believe he is rich?"

Hindolo looked at him, surprised and unsure of what to say.

"The girl T'challa loved fell in love with another and left him. She lives in another village now and has three kids. T'challa was never able to move past that, which is why he lives alone. No kids, no wife. His parents also died a few years ago. He's alone, and he would give all his riches to have a family.

Now, do you understand what it means to be rich? Our riches and success are not the paper in our pockets but the connections we have and the love of those who love us back. There are all sorts of riches, but only you are allowed to determine how successful you are." Explained his father.

The more Hindolo thought of that conversation, the more he felt like crying. His father had taught him the wrong things.

His father had been wrong. While the old concept that money is not ever, money buys a whole of things in the world now Sattuys.

There are no riches except the money in your pocket. That piece of paper holds more power than anything else. Any and all connections and loved ones leave when that piece of paper leaves you. Your entire worth is based on how much money you have, which was a hard lesson Hindolo had to learn when he came to this city.

Adjusting to city life was a huge challenge on its own. Things had been so different in the village. They had nothing to worry about. He had his father, his friends, and the villagers who adored him. In the village, everyone was there for everyone. They were all a family and never left one of their own behind. They would do everything together, and they protected each other.

There was no hesitation in times of need asking for help, be it asking for some missing ingredients from the neighbor, which you need while cooking, or some extra cash when you do not have some to spare at the moment.

Going from that to the city where it was every man for himself and even your own family did not treat you well was a hard thing to adjust to. It's something most adults don't get used to, and in contrast, Hindolo was just a little boy who could have never imagined how much things would change in a year.

He missed his people. He missed the warmth and closeness of being part of a community where people genuinely cared about him. Here, they only cared if they needed something from you, and the moment you were of no use to them, you were discarded.

In the city, life was completely different.

People lived a rather fast-paced life and had time for no one. Most people here worked blue-collar jobs, providing entertainment or providing some service to the entertainment business. They were busy, focused, and occupied in their own businesses, with only one goal in their mind; to be rich and successful. Somewhere along the way on their path to success and riches, they forgot their humanity and lost what goodness they had been born with.

On the other hand, people got hurt while trying hard to be rich. Not everyone made it. Some tried, but the cruel city life and the structure of society only allowed a few to be rich. Most could be middle class, others low-income high earners with hopes and aspirations never dying. Others got frustrated and went into crime and alcoholism' or prostitution for women. All in pursuit of the god-like item in money.

The city's social order was just that – people were busy and did not care about the indigence of others. If, by some misfortune, you weren't wearing suits and marching into giant corporate cesspools, you had done something wrong in your life.

Such was the norm here. You were either from a wealthy family who was granted lands by the British to conduct agricultural activities, or you didn't consign enough focus to your studies and, therefore, weren't able to study at a good university by securing a European scholarship.

If you did not fall into the top three categories, you were thought to be nothing more than a useless bum.

This was the way of life in the city. You were either a corporate slave, unknowingly a victim of capitalism, or a homeless person lying by the pavement, begging for a few bucks.

There was no in-between.

Hindolo had never been exposed to these things, and the disparity both astounded and frightened him. He did not know how to deal with this unknown reality. His father had taught him to treat everyone right, to be kind, caring, and helpful.

His father hadn't known life in the city, for if he had, he would have taught Hindolo to be strong, resilient, and tough enough not to let anyone or anything faze him. Perhaps his father had not wanted his son to learn these things because had he been alive, he would have never let his boy end up in the city or thrown him to the wolves as his aunt had.

Hindolo was perplexed by this strange disparity. The principles instilled in him by his ethical father didn't align with what Hindolo was witnessing in this city. His entire belief system was collapsing.

His father had been utterly wrong, and Hindolo wished to God his father had made him stronger.

He walked along the street and reached the bridge, which overlooked a river. He stared down at the river, watching the water flow steadily but breaking when hit by ships and rocks. There were no bridges in the middle of the city in his village. The towns were flat and rivers outside of the village. Women go to fetch water, and in the process, they gossip. It is the place where all the gossip of the society is heard. It is on the roadside where the hidden lovers meet and play love. It is on the wayside of the river that ambushes were laid by husbands or wives from their cheating partners. There were simply no rivers running through the streets or roads. This also was strange, and even the flow of the river was obstructed by rocks and debris.

Such was Hindolo's life. He had been steady and good, but his father died, and his aunt brought him here, and now everything had gone downhill.

He had never felt this sad before.

His father had trusted his aunt with his protection and care.

Tired of getting complaints about him, she had enrolled him in school. Hindolo had been ecstatic about that. He had wanted to go to school for a while, and his aunt had finally enrolled him. He went there happily because even if he had nothing else, he had the opportunity to study. His father had always wanted him to get a good education and make something of himself but also to never forget his roots and find happiness in everything.

It was a struggle, but he was trying.

All his hopes, however, were crushed once he reached the school.

The schools were in deplorable conditions. The desks were broken, and the chairs were taped together. There was no supervision and no proper education. The teachers had been on strike due to low pay, and those who were not on strike did not care about teaching anything. All they cared about was releasing all their pent-up anger and frustration on students by hitting them if they did not answer a question correctly. The strikes were called "go-slow," which practically meant the teachers would go to school but not teach. The central government had threatened to sack all teachers if they embarked on strike; so they went to school but taught nothing. Students will spend days idling about.

Life felt hopeless, but he couldn't give up.

It was getting late, and he had spent his day wandering around hungrily, thinking about life but not doing much else. He decided to head back to his aunt's place.

However she was, at least she had taken him in when he had nowhere else to go, and she was his father's sister, after all.

He knew there would be nothing to eat once he got home, so he tried to find something to eat here. He didn't have much of an appetite remaining after today.

There was an ice cream vendor nearby. Hindolo rushed to get some ice cream. He fished what little money he had out of his pocket and asked for ice cream. However, he was short, and the vendor refused to sell it to him for cheaper.

"Everything is too darn expensive!" Hindolo sighed in frustration.

There used to be a time when he did not even have to think about money before having ice cream. It was a ritual between him and his father. He had spent nearly every day devouring the sweet taste of caramel ice cream; however, back then, he had been too naïve to comprehend the struggles that awaited him in his near future.

Hindolo turned away and walked back home on an empty stomach, prepared to go through the night with no food in his system once again.

When he reached home, his aunt was sitting at the table smiling gleefully, holding a letter.

"Hi, aunty, what's that?" asked Hindolo politely.

"This is my payment for taking care of you at my expense." Replied his aunt nastily.

"What do you mean?" asked Hindolo incredulously.

"It means, you insolent brat, that I needed money, and now I have it. Why else do you think I took you in?" replied his aunt sadistically.

"That is my money. My father left it for me, for my education so that I would have a good life." Yelled Hindolo.

"Lower your voice, boy. I took you in when no one else did. I deserved this money. It's not my fault your fool of a father trusted me with it." Laughed his aunt.

Hindolo stared at her.

His jaw had dropped.

When Kamoh was alive, Demy was an entirely different person. She was self-centered around him, singing his praises and orchestrating lectures on the importance of family above materialism.

Kamoh was too compassionate to recognize his sister for what she was: a selfish, egotistical liar. Although Kamoh led an honest life and did not amass a fortune, he did set aside money for his son. The sum was insufficient to purchase a home in the major city, but it was plenty to pay for Hindolo's schooling. Before Kamoh died peacefully, Hindolo told him that he would finish his studies with Kamoh's life savings and become an important person in Sierra Leone's history.

When his father died, it had been hard not to lose hope, to keep going on, to keep himself together. He tried to find the good in things like his father always encouraged him to. He tried to find something to hold on to so his mind wouldn't drown him.

That was when he began fantasizing about the education he would receive in Sierra Leone, which would, of course, be far better than any he could receive in his village. His education was his father's main priority, and he was happy he had a chance to fulfill his father's dreams. He always believed that his mother was his guardian angel and that she was watching over him from heaven, but now he had two guardian angels.

This gave him the hope he needed to be optimistic about the life he was headed to.

How could he have ever foreseen that something like this might happen?

He didn't recognize his aunt right now. There was a complete stranger standing in her place.

Who was this person?

Had it always been about the money?

She would visit them twice a year, but now that Hindolo thought about it, while she was really nice to them, she had always asked his father for money on every trip. So perhaps it truly had been about the money. Maybe that was the only reason why she even came to visit them.

She had promised his father that he would have a decent education and live a comfortable life, which is why he had signed over his entire life savings to her. The documents were being processed, and she had brought Hindolo to live with him in case the courts demanded proof that she was acting in Hindolo's best interests. The guardianship had to be officially recognized in order for the funds to be transferred.

This was a necessary measure that the government had to take since child trafficking under the guise of guardianship had become quite common. Children would be brought from the villages and left to struggle because their parents were not there to look after them anymore. Some were sent to sell little things on the street or become beggars, and the worst happened to the girls who were forced into prostitution to service their guardians and often their friends.

The documents had finally been officially released, and she now had complete authority over the funds her brother had left, which meant she probably would not tolerate Hindolo's presence much longer.

Hindolo decided to back down. However she was, whatever she had done, he had a roof over his head. He would study in the government school and still make something of himself.

So this was where his life had led him as a Broken and powerless boy. He fought to hold back the tears, but there was nothing he could do. They slid down his cheeks as a lump formed in his throat. He missed his father more than ever. He had never felt pain like this before, and he tried to be brave, to have faith, but it was too difficult, so he let himself cry his heart out and vowed that no matter what happened, he would not let anything dissuade him from his purpose from now on. If no one else looked after him, he would look after himself.

The next morning, he got up at seven a.m. for school. The school was a twenty-minute walk away. He quickly changed into his clothes, grabbed a book and a pencil, and stepped out.

While walking, his slippers tore.

He came to a standstill to check whether it was manageable, but his shoes had a large split in them. He sighed quietly to himself. He couldn't believe his misfortune. You'd think that after everything he went through the night before, life would become a little easier for him, but his luck just kept getting worse and worse.

He sat on the sidewalk, his head in his hands. He'd have to go barefoot to school since the streets were filled with litter and broken glass. His aunt would not even take him to the doctor if he ended up wounding his feet.

He kept looking around anxiously, trying to figure out what he could do, how he could fix this. Just then, a woman passing by stopped. She was holding a box in her hand, and she gave him a banana and an apple. She smiled at Hindolo and walked away quietly.

Hindolo watched her as she left. As his stomach gurgled, he was grateful for the food he had received but also stunned at what he had become.

So this was where his life had led him—reducing him from an ordinary happy-go-lucky youngster to a pauper surviving on the scraps and crumbs left for him.

"Perhaps this is from my parents, my guardian angels," he thought cynically. "They were aware of my hunger."

That thought was the only source of consolation he had had in days.

That mosque nearby called for prayer just then. He was late for school, but he had an idea.

Since Muslims are not permitted to wear their shoes inside the mosque, they normally leave them outside. He waited for the throngs of people filling up the streets to remove their shoes and enter quickly. When the prayers began, he hurriedly changed into shoes that fit him and dashed to school.

He didn't want to steal, but he didn't have any other options. He couldn't afford food, so shoes were a luxury.

The school was the same as it had been the day before, with teachers who were furious and short-tempered. He devoted all of his attention to his studies, forgetting about everything else going on in his life.

He was a smart boy, and his teachers could see it.

There was only one teacher who was not completely heartless. Once the bell rang for lunchtime and everyone ran to get something to eat from the vendors outside, Hindolo stayed back.

Miss Jasmine realized he had no money to buy food, so she came and made him an offer.

"You can come early in the morning and clean the class every day, and I will, in turn, buy your food; how does that sound?" she asked.

Hindolo, who was surprised, nodded and agreed. He was fine with working for food because it would keep him nourished and allow him to concentrate completely on his studies, and that was all he genuinely cared about.

He had to fulfill his father's dreams and be successful, win or lose, but he could never be the kind of man his father envisioned him to be. Life had changed him too much. Hindolo had realized that there is no richness in having good connections and loved ones around you. He had loved ones and connections; what good did it do him? You don't have everything if you don't have money, and it was time to forget all his father had taught him and become the ruthless, selfish person everyone else in this city was.

CHAPTER 3

CITY LIFE

Hindolo was now painfully aware that this was his life. He could not count on his aunt for anything; she didn't care, and she wouldn't be here, so he expected nothing from her. He figured that even if she couldn't love him as a son or treat him as a member of her family, she would at least treat him like a human being once she realized that Hindolo was not badgering her about the money anymore.

However, her attitude only worsened. She would not give him food, and any time Hindolo was in the same room as her, she would constantly berate him, reminding him how he was a burden on her, how she never wanted him, and how she was grudgingly responsible for a child that was not even hers. If Hindolo, god forbid, had the audacity to say anything back, she would slap him across the face and throw him out for the night.

He would then have no choice but to roam around on the streets, find a quiet spot, and go to sleep there. He would return in the morning to grab his backpack and head to school. His aunt did not care if he lived or died. In fact, she would prefer it if he died since she could have all his money and no accountability.

Due to never having proper food, Hindolo began losing weight. He would either beg for food or steal, but most nights, he would go to sleep hungry.

It wasn't like people in their area were better off. They were all cut from the same cloth of destitution. No one had anything to spare, and what little they did have, they would selfishly keep to themselves.

He would walk to school every day and focus all his energy on working hard, but that was almost always near impossible. Teachers would be on strike nearly every day, but he still looked forward to school because that was the only time of the day when he could have a proper meal, due to Miss Jasmine. He would put in extra effort in cleaning up her room, and he would take up odd jobs every now and then around the school to earn money.

It was never enough, but it was better than nothing.

One night, after school, he went back home. He usually went home late at night only to sleep, but that day, he had a lot of homework and exams to study for, so he came back early.

His aunt had friends over. As soon as he came inside, everyone turned around and looked.

Aunt Demy looked furious.

"Who is this boy, Demy," asked a surprised man.

"Oh, he's just this local beggar who I help every now and then." Replied Aunt Demy quickly as she got up and came to Hindolo.

Hindolo stared.

Nothing his aunt did surprised him after everything she had done, but it would still always sting. His aunt grabbed him by the neck and threw him out.

"I have people over; don't come back tonight." She hissed.

"I have work to do I need the light. I'll be quiet, I promise." Said Hindolo desperately.

"Boy, do I look like I care? Get out." She scoffed and shut the door in his face.

Hindolo stared as they turned the music back up, and he could feel the faint hum of the song. It was one of his and Kamoh's favorite songs. He leaned his forehead against the door and closed his eyes, trying to imagine his dad, but he quickly snapped his eyes open.

It didn't matter what his father would think of all this, for he was no longer alive, and all Hindolo had was himself. He glared at the door, turned his back, and left.

He wandered around on the streets, unsure of what to do and where to go. How could he study now? There were no libraries, and his friends had the same situations in their own homes, so he couldn't go there and risk another humiliation. He decided to walk to the other side of their area where the streetlights worked.

The streetlights were dim and yellow, but they were better than nothing at all. He sat down on the curb and took out his books and a pen. Studying was the only thing that helped distract him from the harsh realities of his life.

The street was quiet. Aside from a few homeless men asleep here and there and a few druggies smoking or injecting themselves, there was no one here.

A stray dog just then came up to Hindolo and gently nudged him. Hindolo patted him on his head and stroked him. The dog whined and laid his head on Hindolo's lap as Hindolo continued to stroke him.

"You haven't been a pet in a long time, have you?" asked Hindolo.

The dog whined again.

"Me too, buddy. I guess we're both alone in this world. No mama, no papa, no family; no one who cares. We only have ourselves." Sighed Hindolo.

He knew he was talking to an animal, but he couldn't help himself. He was just as starved for affection, love, or even a kind word as this dog was.

He looked around at the city in front of him. It was very different from laws. The air was heavy and thick, and there were no trees or greenery. The sidewalk was rough and littered, and the light was so dim it strained his eyes to study. He did not have a choice, though. He had to continue studying in this light or suffer the consequences.

Once he had finished his studies, he decided to go to sleep. He did not even know the time as there were no watches around them, and he did not own one himself, but he could tell it was late.

He grabbed his things, stuffed them in the back, and began walking back to where he lived. There was a spot there on top of one of the houses where he had set up a place to sleep on days like these where he had nowhere else to go.

The roofs of most of the houses were made of tin. He had found an old blanket, and he had hidden it here in a box. He took it out and laid it underneath himself. He then got on his back and stared at the open sky. It was still uncomfortable and hurt his back, but it was better than nothing and much safer than the street, where it was easy to get robbed.

The sky was dark, and there were a few stars here and there. Every time Hindolo slept here, he would look for the brightest two stars and imagine they were his parents. He would talk to those stars, imagining they could hear him and say good night before going to sleep.

Despite excelling in school in all his subjects, he could not make any friends. After all, while no one here was rich, most kids were not homeless and destitute like he was. They wouldn't invite him to go out and eat since he never had any money to afford food. He couldn't go to birthday parties since he had no clothes to wear, and the ones he did have were either dirty or had holes in them. He did not even have proper shoes. The ones he wore were those he had stolen, and even they were starting to get worn out.

He could rely on his aunt for neither shelter nor food, which meant he needed a stable source of income. There was an old man in the area who worked as a cobbler with whom Hindolo had grown acquainted.

The old man had seen Hindolo steal apples once and had let Hindolo hide with him as the police chased him. He knew how cruel the police could be to young boys. Hindolo had sat next to him, pretending to be his apprentice.

The police did not even stop; they kept running.

"I know you don't want to steal. You have no choice but don't make a habit of it. Try to find honest work." The man had said.

"Who would give me work, and what could I possibly do?" asked Hindolo.

"You could work as a cobbler." Suggested the old man.

Hindolo nodded and replied, "I could, but I do not know the first thing about cobbling."

"I will teach you. It is not that hard. Besides, this way you will have enough to feed yourself at least." Smiled the man, "Life gets tough for all of us, child, but we have to fight through it."

The man then took out one of the apples Hindolo had stolen and chuckled, "This is my payment."

Hindolo smiled and nodded.

The man had then explained how cobbling works and taught him a few tricks. Hindolo was a quick learner. It did not take him long to absorb whatever was being taught. The cobbler even let him practice on the next customer who came, and he was surprised to see how well of a job he had done.

"I'll tell you what. How about you come here after school every day. You can study and earn and then go back home with some extra money and a meal." Proposed the man.

Hindolo was taken aback by this offer. This was the second time anyone had shown him any sort of kindness in this city where no one had anyone's back.

Hindolo nodded and readily agreed.

This became his life.

From going to school to the cobbler, studying in between breaks. If his aunt would let him in, he would sleep there; otherwise, he would go and hang out with his friends.

The older they got, the harder their lives became, and that is when they discovered marijuana. Most of his friends had resigned themselves to their fate. They did not see a life beyond the slums for themselves. They did not care about their studies, and since that was the only way out, they had nothing more in store for the future than they had right now.

They began smoking marijuana every night. Hindolo joined them in doing so, but he did his best to control his habit and not let it become an addiction. They would often stay up all night talking about life and how different things could have been had they been dealt a fair hand in life.

They would indulge in gambling as well. This was something Hindolo stayed far away from since he hardly ever had money in his pockets, and what little money he did have, he had to save since he had no other way of getting money. He would use all the spare cash he had to buy books and pens. It was tempting at times to watch his friends earn double what they paid, but when they would lose everything, it would be a swift reminder that he was doing the right thing. Regardless, he would still accompany his friends wherever they went.

Despite all this, he pushed through everything. He studied hard and continued to excel in both his studies as well as a cobbler. People loved his work and would often come specifically to get their shoes mended by him. His teachers were very happy with how well he was doing in school. He was the least problematic child, according to them. They didn't have to put in any effort to teach him, yet they received all the credit any time he did well.

His aunt's behavior stayed as bad as always, but Hindolo did not care. He only went home to sleep, and if she let him in, he would sleep; if she didn't, he never argued and would simply go on his way. All he cared about was finishing his education and moving far away from this place and ending this nightmare once and for all.

CHAPTER 4

UNIVERSITIES

After working long and hard for years while still suffering at the hands of his cruel aunt and fate, which seemed to keep testing him, Emmanuel was finally able to get the ticket he needed to get out of the dreadful life that he had grown accustomed to living.

He had practically lived on the streets, considering the fact that his aunt only let him in when she was in the mood to do so. He had worked hard to provide for himself day and night, doing one thing or the other, working as a cobbler after school and sitting on the streets after work to study under the streetlights. He was always one of the top students and remained that way throughout his teen years.

Finally, it was time for his board exams. He needed exceptionally good grades in order to not only get admission to a good university but also to secure a scholarship. He had spent the entire week outside different restaurants, hotels, shops, wherever he could find light and study there.

His aunt had invited him back home during his exams surprisingly. Hindolo didn't care, though. He knew she hadn't invited him back because she had suddenly begun caring for him. The only reason he was invited back was because if he ended up getting a good enough grade, he would leave forever, and his aunt would never have to worry about him again.

Hindolo worked hard day and night and took his exams. When the scores came out, he had not only topped the entire district but also the entire city. He had also been offered a scholarship at the University of Sierra Leone, which was a huge honor since it was the oldest university in all of Africa.

Hindolo was overjoyed. He had worked hard the last six years and he was finally getting rewarded for it. His friends were even more excited than him. They had resigned themselves to the designs of fate, but they had never let Hindolo give up, no matter how hard the days got. They had high hopes for him, and they never lost faith. On some days, it was only their hope that had kept him going and it had all paid off. He didn't even bother going to his aunt. He showed Kgosi, the cobbler who had helped him all those years ago and who had allowed Hindolo to work for him so he could provide for himself. Kgosi was thrilled. He gave Hindolo a big hug and congratulated him while wishing him good luck.

Hindolo then headed to his old school to both thank and show Miss Viola his result. She was so happy she began to cry and she told him repeatedly how she had always believed in him. She helped Hindolo fill out his forms there and then paid for the stamps so he could post them. Despite Hindolo's protesting, she still insisted.

Once the letter was posted, there really wasn't much more left to do. He slipped into his aunt's house after that to gather his things. He always used the back entrance, climbing through the window. He headed to his room and grabbed what little possessions he had. Mostly, he just needed to get the pictures of his mother and father and of the village. Everything else he stored in his friend's house.

Once everything had been packed and was ready to go, he decided to leave through the front door. He had no intention of ever coming back, so why did it matter if she got mad at him. What would she do? Kick him out? She was more than welcome to do so since, for the first time in years, he had someplace to go.

He headed out of his room boldly and out into the lounge towards the main door. She had guests over, and while she usually kicked him out, today, she called him in.

"This is my nephew, Hindolo. He just got accepted to the University of Sierra Leone on full scholarship." She boasted, beaming with pride.

The guests murmured in appreciation as his aunt took it in and smiled at him. Suddenly a fire erupted in him, and he allowed himself one final moment of defiance before vowing in his heart to never see her again.

"I got here due to my own hard work. All you did was steal the money my father left me and reduce me to the rank of a beggar. This success is mine and mine alone you greedy woman. You did not even give me shelter or food. Don't try to act like you had any hand in this." Said Hindolo defiantly.

His aunt's face turned red as the guests whispered in shocked voices.

"How dare you, you insolent boy-" she sputtered.

Hindolo held up a hand and replied, "Enough. I've had it with your lies. I do not need to stick around for them any longer. This is goodbye, aunt. I do not wish you well, and I hope I never have to see you again."

He watched in satisfaction as his aunt continued to sputter, and her guests stared in confusion and disgust. He had never felt this happy in his life. Even his good results did not bring him as much joy as telling his aunt off did. He left her house and headed to meet his friends.

Hindolo had a train ticket booked for the next day. His friends and he ended up going to their childhood spot where they used to play war games, only to discover that it had all been grazed down. The forest, the beautiful plants, the animals, they were all gone. All that was left behind were high-rise buildings and a concrete jungle.

The friends had been so busy in their lives that they hardly had the time to come out to the side of the town where the forest was. It was devastating to see what had become of the place that had stored the few good memories they had of their childhood.

The development had destroyed the beauty of the area. There was no greenery in sight anymore. Everything had been taken down and destroyed. It was all dreary, drab, and gray. It was hideous and a rather crude vision of what most of the world would look like someday.

They spent the night reminiscing, talking about their childhoods, recalling every little thing, all the times they got in trouble, all the fun games they used to play. Hindolo knew in his heart that no matter where he went or what he did, he would always miss his friends a lot.

The next day, they all accompanied him to the station and wished him luck, making him promise he would visit them someday.

As Hindolo got onto the train he couldn't help but feel excited for the future that awaited him. Once again, he was about to start a new life, and he hoped that this time would be better than his last as he took his seat and daydreamed about his future.

The train ride was shorter than he expected, or perhaps it went by sooner than one would expect since he was too busy thinking about his future the entire way. Once the train came to a halt, he got out. His papers were secure in his bag, but he double-checked just to be sure.

The University of Sierra Leone was in Freetown, which was the capital. Hindolo asked for directions and discovered the university was a twenty-minute walk from the train station, so he decided to save his money and walk to the university. He had limited resources, and he had to be careful in how he utilized them until he had an alternate source of income.

He walked all the way to university, taking in the sights and learning new ways around town. Once he reached the university, he headed to the reception. They asked him for his papers which he provided, and they then asked him to fill in more papers. He obliged, and soon enough, they directed him to the dorms where he would be staying with two other guys.

The dorm was disgusting, at the very least. The bed was tiny, and the mattresses were rotten. There were two dingy lights in the room and one fan, which was so slow you could hardly feel the air coming from it.

His roommates were already in the room when he got there. They seemed like they came from average middle-class families judging from their suitcases and clothes. They wore the same jacket and smiled at him as he came in.

49

"Welcome to our room, brother. I'm Abidemi, and this is Ade. We're in the Bai Bureh Hall fraternity. If you need anything, do let us know."

Hindolo was surprised. He was mildly taken aback by their openness and sweetness. Life had taught him that no one was nice to you without an ulterior motive, and he had doubts as to why they were being so good to him. However, he decided to put those doubts to bed for now, considering he was in a new place with no friends, and he did not need any enemies.

He smiled politely and nodded. He set his stuff aside on the one bed, which was empty which did not take very long considering he had such little stuff.

"We're going out to explore the campus. Would you like to come with us?" asked Ade.

Hindolo shook his head and replied, "I'm very tired right now. Perhaps another day?"

"Of course!" He replied and exited the room with his friend.

Once they were gone, Hindolo decided to go out as well. He had been given his class schedule, and he could see he had a class today as well. He headed out of the dorms and the building, walking to different places to see the stalls for different clubs and fraternities, encouraging students to join them.

He found a shortcut to the other side of campus behind a building and decided to take it. The shortcut had a sharp turn to it. As he approached the cut, he heard voices. He stopped and decided to listen before proceeding.

"Why has he not joined us yet?" Asked a male voice.

"He joined the sigma's." Replied a voice.

"Now, wasn't that your job to ensure that this does not happen?" asked the voice in a bored tone.

"Yes, but they got to him first. They threatened him to join. He was too scared." Replied another voice.

"And you couldn't scare a freshie? What do you have a knife for?"

"I'm sorry, I'll recruit another one right now. There's a ton of freshies coming in today. Just give me another chance."

There was a sigh.

"We will give you another chance, but you have to pay the price."

There was a movement of feet and a swoosh, and Hindolo heard muffled screaming. He took a step back, scared and petrified.

"Don't disappoint us again, or we will do worse than this, understood?"

There was muffled crying.

This was enough for Hindolo, and he ran out of the alley back onto the main road of campus. His heart was beating in his mouth. How was university any different than the streets if you could get stabbed here by anyone at any time and be able to do nothing about it?

He continued to walk through campus with his head down, working hard to keep his face as expressionless as possible. He finally reached the building where he had his class and went inside. The class was in the auditorium which was packed.

The professor then came into class. He told us all to take a seat. I was excited. This course was about philosophy, and the lecture focused on motivation. Hindolo sat in his seat excitedly, hoping for an exciting lecture. However, the second the professor began speaking, he realized that this place would not be living up to his expectations. In fact, it would be worse than public school.

The professor was a bitter man whose form of motivation was to tell students that they would fail. He spent an hour-and-a-half-long class tirelessly berating Africans and how we will never be as good or as successful as others. He even told us how our degrees were worth nothing and most of us would fail in life. A few handfuls might succeed, but it was a big maybe, in his words.

Hindolo couldn't believe his ears. He couldn't believe that a teacher could be this terrible to his own students and teach them such things that would surely ruin their self-esteem and confidence.

Once the lecture was over, Hindolo walked out of the hall quietly. He was rather surprised and sad at the state of this university. They had no proper facilities. The chairs were broken, teachers were subpar, and students were attacking one another on campus, and no one took any action. He knew he had to be very careful here and keep his head down now, and no matter what happened, he had to stay out of trouble.

He returned to his room after class to lie down for a bit and ended up falling asleep. When he woke up, his roommates were in the room.

"Hey, you're up." Grinned Ade.

"Yeah," replied Hindolo, rubbing his eyes.

"Okay, come on, we're going out. We'll show you the recruitment process for the fraternities and how they swear in new recruits," said Abidemi.

Hindolo opened his mouth to protest but then thought better of it and decided to go with them. They took him out of the building to another part of campus. There was another building that presumably housed some fraternity. They walked in and stopped at a random door.

Ade knocked twice and then made a strange noise.

The door opened as soon as he did that.

Inside the room, there were a bunch of different guys. Most looked anxious and weary, and Hindolo could tell they were freshers like him. The rest were laughing and looked comfortable, so of course, they were the ones who ran the fraternity.

"Is everyone here?" asked a guy who was standing on top of a table.

Ade and Abidemi were in a corner speaking to a guy, and all three of them were looking at him and eyeing him as if he were fresh meat.

He felt uneasy, but he couldn't say anything.

The man then continued to speak, "Now, if you're all here, then that means you want to join our fraternity. To join us, you have to go through a series of obstacles, and you will have to do what we tell you to, or you will have to face our wrath. There is no turning back now."

Hindolo looked around frantically for Ade. He had not signed up to join the fraternity, but this was why they had brought him in.

"you're mine, pretty boy," whispered a voice in his ear.

Hindolo jumped back.

There was a guy staring at him, and Hindolo had never felt his uncomfortable in his life. What was this man doing?

"What the hell do you think you're doing?" he asked angrily.

"What everyone else is," smiled the man, motioning around the room.

Hindolo looked around to see the freshmen serving the other guys. He felt disgusted. He pushed the man away.

"It is never going to happen."

The man growled at him, and people around the room began edging towards him. Hindolo did not wait for them to speak or do anything. He merely turned around.

And just like that, he ran out of the room and out of the building. He had never run like this in his life before. He ran across campus into his room and grabbed his things. He would change rooms in the morning. For now, he could sleep anywhere. He was used to sleeping on the streets, what was one more night of doing more of the same. It was better than being assaulted or harmed like that guy in the alley.

CHAPTER 5

CIVIL WAR

As time passed and Hindolo tried to adjust to his life at university, he decided to start working part-time. University had not lived up to his expectations, and after what took place in the dorms, he did his best to stay as far away from campus for as long as possible.

Those guys mostly left him alone, but Hindolo had learned enough living on the streets to know not to let his guard down. People always attacked when you least expected them to. He started working at a stationery store a few minutes walk from university. The owner was a gruff man, but the shop had cooled, and Hindolo could study if no clients needed to be attended to. The money wasn't bad, and it helped him avoid campus.

This became his routine.

Getting up, going to lectures, heading to work, remaining at the shop until it closed, returning to the dorm, and falling asleep right away. Hindolo was always too overwhelmed to focus on anything else at first, but as time went by, he became used to his hectic routine, and his mind often strayed. He couldn't stop thinking about how he would sometimes see kids whose parents, relatives, or friends had come to visit them simply because they were missing them. Hindolo had no surviving family. His parents had died, and his aunt was as good as dead to him as well. He didn't expect his friends to come, given that they had their own problems to cope with and the place they lived in was a survival of the fittest. Even one day away could mean someone staking their claim on your territory.

Even after years of living in the city, he had not gotten used to how awful the urban truly was. He missed his village, its people, the river, life, and the pure and fresh air. He hadn't returned since his aunt had hauled him away, and he had never visited his father's grave since the funeral. His aunt had promised to bring him back at regular intervals to visit his father's grave, but this, like everything else she had said, had been a lie. Hindolo had been feeling lonely for a while, especially after seeing other students with their families, and since he had nowhere else to go, he decided to start saving money to visit his village during summer break.

The more he thought about visiting in the summers, the happier he got. He couldn't believe he would see the village again, his old house, the square where he would go with his father, his friends, or any of the other things.

That night was the first time in a long while that Hindolo had found something that had been missing all these years in his life. He found hope. That night, he slept like a baby, soundlessly and peacefully, excited about what was to come. What he could not have anticipated was that his world would soon come crashing down, for Sierra Leone had entered civil war.

The country had been struggling ever since Joseph Momoh came into power, with millions of people starving and struggling to survive as the economy collapsed and the government failed to help its people. Schools shut down after the government failed to pay its staff their wages, and state administration offices came to a standstill due to a lack of funds. The politics of the country took precedence over the needs of its people, so while millions perished, the leaders squabbled behind closed doors without a care in the world. Hindolo, however, had always been so caught up in his own matters that he had no time to focus on politics. His priority was surviving, and then he would worry about who was taking over where.

Life continued to carry on as normal for most people, Hindolo who could do nothing but try to survive just like him. All they could do was ignore it until it couldn't be ignored. At that moment, ignorance truly felt like bliss. However, the country's problems were exacerbated in March 1991 when unrest from Liberia surged over the border into Sierra Leone. President Momoh reacted by sending soldiers to the border area to fight an assault by Liberian rebels who were commanded by Charles Taylor's National Patriotic Front of Liberia.

Sierra Leone's army was found to be at a disadvantage, and they were attacked not just by the NPFL but by the Revolutionary United Front, commanded by veteran Sierra Leone army corporal Foday Sankoh, who had been coordinating with the Liberian rebels. He had extended support from Charles Taylor, which marked the start of a protracted and terrible civil war that would destroy the lives of the people of Sierra Leone.

Momoh was then, however, removed in a coup that had been orchestrated by Capt. Valentine E.M. Strasser claimed that the harsh circumstances that had been faced by the troops who had combated against the insurgents were one of the grounds for overthrowing Momoh. This was a reason which had been plastered all over the channels and the newspapers.

The shop where Hindolo had worked before going to university, was directly in front of a cafeteria where there was a large screen. Hindolo received most news updates when he had a spare moment to focus. That day, when the coup came, most people stopped what they were doing to watch the broadcast. Hindolo had been able to watch the broadcast live. The uncertainty of what would happen next was the talk of the town. Plenty of people in the area had begun to flee. They were, of course, the wealthier ones who had somewhere to go. They warned everyone that a civil war was approaching, which would ruin them all, and it was best to flee as far away as possible.

Some people did not take them seriously. Those who did listen to them either fled themselves or had nowhere to go. After all, where could they possibly go? They had no money, no visa, no immigration to other countries. All they could do was sit back and pray that things did not get worse.

Strasser was later appointed as the head of the National Provisional Ruling Council (NPRC). During Strasser's presidency, the RUF expanded their dominance over different areas, expanding their bases and terrorizing more people into obeying them. They took over villages, towns, properties, and businesses, most notably that of the valuable diamond mines. This is where the "blood" or "conflict" diamonds originate from which were being used to pay for their militia operations.

In March of 1993, the Economic Community of West African States Monitoring Group (ECOMOG) sent primarily Nigerian soldiers to Freetown to help the Sierra Leone Government reclaim the diamond belts and drive the RUF toward the Sierra Leone-Liberia frontier.

These things kept happening, but Sierra Leone had remained untouched thus far. Hindolo would get updates regularly from the

TV, which was in the shop where he worked, and he worried every day about what would happen next. A lot of professors and students had begun to flee, but most had stayed because they had nowhere to go.

The Sierra Leone state then recruited a South African-based paramilitary company in March 1995 to eventually destroy the RUF.

While the war had been ensuing, no one had been allowed to leave the city and no one wanted to as well. Every day, however, someone or the other receives a letter or telegram providing updates. Entire villages were being burnt to the ground, leaving behind no survivors. People were being hacked left and right, and no one could do anything about it. Hindolo began getting scared for his village, especially when a girl who was in his university received a letter informing her of her village being burnt. Upon inquiry, Hindolo discovered her university had only been a short distance from his own.

She had been too hysterical to say much else, so people left her alone except for some of her friends, who stayed by her side trying to comfort her. Once she had calmed down, Hindolo approached her again to ask about his own village. The girl, Ahosi, was white as a sheet. Her lips were quivering, and her eyes were wide as she met his gaze and handed him the letter.

Hindolo took the letter and scanned it with his eyes. Sure enough, Ahosi's father had survived the fire and sent her a letter informing her of the villages that had been burnt and to not come looking for her. With each village name, he had marked whether anyone survived or not.

Hindolo's heart dropped to his stomach. In front of one village, it read, 'Five survivors.' In front of others, it read, 'No survivors, everyone dead.'

He kept reading as quickly as he could, praying with all his heart that his village was one of the lucky ones. To his dismay, however, his village had been set on fire, and the description said, 'No Survivors.'

The letter fell from Hindolo's hand as he stood there in shock. They were all gone. He did not even know if his parents' graves had managed to live through a battle that was not even their own.

He doubted it.

He could feel Ahosi and her friends' eyes on him as tears welled up in his own. He couldn't believe they'd all vanished. He hastily excused himself and started heading to the opposite end of campus. His pace picked up somewhere along the route and he kept running until he couldn't anymore, and as he wept, he let it all out. He cried uncontrollably, unable to conceal his agony or sorrow any longer. He'd already lost so much, and now his home was gone, too. He genuinely had no one left. He had absolutely nothing. He was now utterly alone.

After that, what could he have done other than stay normal? So that is exactly what he did. He would go to work, stay on campus, and walk around, reminiscing and wondering what things could have been like. That fire hadn't just burnt his village down. It had engulfed everything Hindolo had ever held dear to himself with its flames until nothing remained but ash. He carried the pain of it with him every single day, as did most people who had lost people, too. Every day, there were news, letters, messages, and information about people killed, maimed, tortured, or taken.

Later, however, Sierra Leone was re-elected as a democracy in March 1996. This was a huge victory for its people, who had been suffering under the fear of what would happen if the government failed. There were celebrations all over the state as the receding RUF signed the Abidjan Peace Treaty, effectively ending the hostilities and putting an end to the war, at least on paper.

However, they should have known better, for in May 1997, a band of Sierra Leone Military commanders launched a coup and created the Armed Forces Revolutionary Council (AFRC) as the state's transitional government. They encouraged the RUF to rejoin them, and the two forces now dominated the nation's capital, Freetown, with minimal opposition.

The conflict was proclaimed over by Johnny Paul Koroma's new administration. Nonetheless, robbery, rape, and slaughter, primarily by RUF soldiers, immediately followed the interim administration's proclamation, demonstrating its vulnerability. ECOMOG soldiers came back and took back Freetown at the behest of the Koroma administration but could not secure the surrounding areas. The RUF kept the civil war going.

The country was now completely devastated. The RUF was fast approaching where Hindolo lived, and he had had enough. He decided that he would not die without putting up a good fight against these savages.

As a student in school, he did not have the resources to purchase weapons, but he could join those who could, and so he joined others to oppose the overthrow of the democratic government. The students plotted that they would not stay quiet in the face of this oppression. Hindolo began organizing a fight against the invading RUF with the students. They took whatever they could find that could potentially be used as a weapon.

The RUF got to them before they could invade, and they assaulted everyone engaged in the operation to take down the RUF with AK47 weapons and real machetes. As they attempted to flee, the campus descended into anarchy. People were scrambling all over the place. As Hindolo ran for his life, he could see the RUF shooting their weapons, yelling, and savoring the misery and despair on people's faces in every direction. A girl fell behind Hindolo. As Hindolo escaped, he turned to see if she was alright, but all he witnessed was a soldier stab her with his machete as he pulled her away, taking down her trousers as she cried in anguish and to preserve her virtue, but it was all in vain. These men were savages they did not care.

He had helped plan the demonstration, and he barely escaped the clutches of death.

He ran for his life to Guinea Conakry and lived in squalor for ten months. For democracy, he fought and ran. He could not live in a state that was being run by terrorists. Poverty was preferable to that. Where he had lived half his life as a pauper, what was a few more years for the right reasons. He often wondered how quickly things changed. He had just started making something of himself, and even after all that hard work, he was reduced to worse than his condition previously.

Time passed by, and eventually, democracy was restored. Hindolo returned back to the city and began helping the community restore itself. He found an apartment near the state house, and he worked in a cafeteria during the day. The nightmares, however, were far from over.

On Jan 6th, the rebels 1999 invaded the capital city, and the city wept tears of blood. Maiming civilians and burning homes. The Revolutionary United Front (RUF) rebels launched an offensive against the Sierra Leonean capital, Freetown, at dawn on January 6, 1999, seizing it from government forces and army personnel of the Nigerian-led joint force known as ECOMOG, the Economic Community of West African States (ECOWAS) Cease-fire Monitoring Group. The fight for Freetown and the subsequent three-week rebel invasion of the capital were characterized by the methodical and pervasive abuse of all classes of heinous acts committed against the innocent civilians of above one million people and denoted the most rigorous and condensed era of infringements of human rights in Sierra Leone's eight-year-long bloody conflict.

The rebel soldiers chose to not distinguish between civilians and militants and targeted whoever they felt like harming. It became a thrill for them to murder, pillage, rape, and loot. The rebel soldiers in Sierra Leone deliberately killed, tortured, and raped people during their January onslaught. Entire families were shot to death on the street, both children and adults had their limbs chopped off with hatchets, and teenage girls and women were kidnapped and taken to rebel strongholds where they would be sexually tortured for hours on end, often kept there as slaves. It wasn't like the state or supporting troops were better. They were all the same. All of them committed terrible atrocities against the people of Sierra Leone, and there was no accountability. The civilians were merely fresh meat for them to do as they wished.

Young females under the age of seventeen, especially those considered virgins, were intentionally sought. Hundreds of them were later taken by the rebels permanently and never heard of again.

The rebels perpetrated many acts of torture, including amputations of hands, limbs, legs, as well as other parts of the body. Several hundred individuals, predominantly males but including women and children, were wounded and killed in this manner in Freetown. Twenty-six citizens had their arms amputated twice, while others were tortured.

Despite having access to women captured to be used as sex slaves or fighters, the RUF often assaulted non-combatants. The militiamen also cut the RUF initials onto women's bodies, putting them in danger of being misidentified as enemy fighters if they were apprehended by government troops and thus raped again.

Women in the RUF were required to give sexual favors to the militia's male members. Very few women had not been victimized by sexual assault on many occasions; gang rapes and solitary rapes were widespread. The RUF was responsible for 93% of sexual assaults carried out throughout the war. The RUF was known for civil rights crimes, often amputating victims' limbs and legs. Government military men and militia abduction of women and children for use as sex slaves were also just as expected and widespread.

Throughout Sierra Leone's quarter-century civil war, women were kidnapped into refugee camps and made into slaves.

Tens of thousands of people were raped and murdered. Over 5,000 juvenile soldiers were enlisted forcefully, and many were drugged and forced to perform atrocities. Most notably, the RUF became renowned for widespread mutilation. The rebels also severed the legs of about 20,000 victims.

They would go village after village, torturing, raping, looting, killing civilians, and then setting everything on fire, reducing entire districts to ashes and leaving hundreds of thousands of people injured, dead, and homeless.

Hindolo's house was closer to the State House and was the first point of attack. They would kill people in the middle of the street and call it Operation No Living Thing.

Hindolo was forced to dance on the street with the words "We want peace," while the rebels mocked the scream of thousands of people forced to dance on the street with the words "We want Kaka." Human shields were employed against the EXOMOG fighter planes. The rebel commander who had captured Hindolo's home against him was listening to the radio and explaining to the rebels what was happening since most of the soldiers were ignorant, drugged-out youngsters with firearms. With no school or university, life was upended. People faced hunger and destitution as regular life ceased to exist.

Hindolo lived through all of this. He survived once more.

CHAPTER 6

BELIEFS AND IDEOLOGY

The war had ravaged the country, leaving no corner untouched, no one unaffected. Those unfortunate souls who were unable to flee were trapped in a never-ending nightmare, caught between the state and the insurgents. Both sides claimed to fight for the people, to be their saviors and their protectors, but it was the people who suffered the most at the hands of their self-proclaimed guardians.

When the conflict had first begun, perhaps then it had been noble, a righteous cause erected to end their oppression, to free the people. But now it was as if even God had turned his back on them, and they were left to suffer at the hands of the devil himself. The country was in a lawless state, a land where justice was a forgotten dream, where peace was a myth. The population was merely pawns in the game of the state, and the insurgents were herded and slaughtered with no regard for their lives, no one to stand up for them, and no one to defend them.

It was a tragedy of epic proportions, where hope was, but a distant memory, and survival was a daily struggle. People continued to face unimaginable atrocities as they fought to survive, their lives hanging by a thread in a world without mercy, with no one to turn to, no one to save them.

As the war raged on, the economy of the country was dealt a devastating blow. Insurgents seized control of factories and warehouses, bringing production to a screeching halt. The few goods that managed to reach the market were met with exorbitant price tags as producers capitalized on the desperation of the people.

The people of the country were living in a living hell. They were trapped in a cycle of violence and poverty with no end in sight. They were desperate for peace, but it seemed like an impossible dream.

The most basic necessities, such as food, became luxuries that only the wealthy could afford. For the average person, the cost of a single meal was equivalent to a week's salary that they would have earned had things been normal. The harsh reality of starvation and poverty loomed over the nation as the once-thriving economy lay in ruins.

The people were left with no source of income, as the war had robbed them of their livelihoods. They were trapped in a cruel cycle of poverty and hunger, with no hopes of salvation. The country was on the brink of collapse, its people suffering in silence as the world turned a blind eye and the war raged on.

Hindolo was now on the run.

When the insurgents had descended upon his town, he had found himself right in the middle of the pandemonium that had ensued. With his life hanging in the balance, survival was the only thing on Hindolo's mind. He gathered what little he could carry in a knapsack and fled for safety. As soon as he had come out onto the street, he saw the men running with their weapons out.

As Hindolo ran, he could hear the screams of his fellow townspeople echoing through the streets. The insurgents had invaded their homes, their wild eyes glinting with a dangerous eagerness. They showed no mercy as they waved their weapons, stabbing and hacking their way through the defenseless crowds whose screams and pleas fell on deaf ears.

The rebels seemed to take pleasure in the destruction they brought. They set fire to the state house, the flames licking hungrily at the walls and consuming the structure with its flames. Everywhere they went, they poured gasoline and locked people inside their homes, creating a chain reaction of destruction and death.

The insurgents were like madmen, consumed by their own bloodlust and high on the drugs that fuelled their violence. They were a force of destruction, leaving nothing but charred ruins and devastation. Hindolo knew that if he did not flee, he, too, would be incinerated.

As the town was laid to siege and the insurgents began mindlessly attacking and killing people, looting homes, and setting stores ablaze, Hindolo hid in hopes of waiting them out and then making a run for it. He was able to get a clearer look at the rebels from his vantage point. As they set ruin to everything their hands touched, he saw their eyes. They were wild like a beast, and their hair was matted and tangled as if they had been possessed by some wicked force.

Hindolo could clearly see that the militants were not in their right minds. They moved like animals, wild and untamed on the hunt for prey, devoid of any feelings or reason, unable to feel the pain that gripped Hindolo's heart when they attacked a woman and her screams pierced through the loud chaos. Drugs distorted their minds, turning them into mindless pawns in the hands of their masters, who controlled them and made them do their dirty bidding.

Hindolo saw the menace in their eyes. It was the type of threat that couldn't be reasoned with or placated. These were not men with whom one could bargain or make a deal. They were numb to the realities around them, a chaotic force driven by their own violent inclinations and the whims of their leaders.

Hindolo knew that he had to get out of there, but how? He was trapped, surrounded by enemies on all sides. He could hear them getting closer, their shouts and screams filling the air. He knew that it was only a matter of time before they found him.

He had to act now.

Hindolo took a deep breath and ran. He ran as fast as he could, his heart pounding in his chest. He could hear the insurgents getting closer, their footsteps pounding on the ground behind him. He didn't dare look back. He just kept running, putting one foot in front of the other.

He ran until he couldn't run anymore. He ran until he thought his lungs would burst. He ran until he could no longer hear the insurgents behind him. He was now safe.

Hindolo had escaped, but he knew that he would never be the same.

He had seen too much and experienced too much. He had seen the horrors of war, the depravity of man. He had seen the worst that humanity had to offer. But he had also seen the best.

He had seen the courage and resilience of the human spirit. He had seen the strength of the human heart. He had seen the power of love, and he knew that these things would always be with him, even in the darkest of times. Hindolo was a survivor, and he would never give up hope.

As Hindolo fled for his life, he knew he wasn't merely escaping the insurgents but also the madness that had consumed them. Praying for his safe deliverance from their clutches, he sought refuge in this land of madness.

Nightfall blanketed the town with a daunting silence. The militants, having already satisfied their bloodlust, had left a trail of devastation and shattered lives behind. However, their insatiable thirst for violence persisted. Under the cover of darkness, they prowled the streets, hunting for their next victims—the few remaining women who had managed to evade their grasp. Their echoes of shouts and moans reverberated through the desolate streets, forming a perverse symphony of ecstasy and anguish, a striking contrast to the lives the insurgents and the people of Sierra Leone now endured.

A small group of insurgents maintained a vigilant watch, scanning the shadows for any signs of danger. Hindolo realized that blending in with these militants was his last hope for survival. With trepidation, he donned one of their masks and seized a firearm, his heart pounding with fear. Navigating through the darkness, he remained alert, sliding past the guards and escaping the town.

Engaging in this perilous charade was his only option. He had to maintain the deception until he was far enough away from the town to discard his disguise and seek safety. His heart hammered in his chest as he cast a final gaze upon the charred ruins before vanishing into the night, leaving behind the horrors he had witnessed.

Once beyond the town's borders, Hindolo discarded his disguise. The unforgiving terrain stretched out before him like an endless wasteland, its rocky crags and barren valleys serving as harsh reminders of the devastation that had befallen his homeland.

Aware that he couldn't risk taking the main road—too dangerous and exposed—he opted to traverse the unforgiving forest, moving with quiet determination that had thus far preserved his life.

During the nights, he slept with one eye open, his hand tightly gripping a knife. He had learned through bitter experience that danger lurked around every corner in this lawless land.

Just a few weeks ago, before his town was invaded, Hindolo had received word that his surviving relatives had fled to Guinea and Liberia in hopes of finding safety. However, he had yet to receive any further communication and remained uncertain of their fate.

But he couldn't afford to dwell on it. His own survival hung in the balance. With insufficient funds to secure food or shelter, each step he took pushed him closer to catastrophe.

Deciding to return to the city where he had resided with his aunt, he hoped that some of his friends had managed to survive. These were resilient men, and he believed they wouldn't succumb easily. Perhaps they could help him determine his next course of action.

Eventually, he reached the city. The deafening silence enveloped Hindolo as he traversed the abandoned streets, his eyes scanning for signs of life. Only the echo of his footsteps interrupted the stillness. He trod cautiously, opting for side alleys, fearful of lurking militants. Memories flooded his mind as he passed by familiar places where he and his friends had once reveled in chaos, tormenting those who lived there. A grin spread across his face as he recalled their escapades of evading the police after being caught playing football—an echo of better days before the ravages of war.

Finally, he arrived at his former apartment building. As he pushed open the door, it emitted a loud creak, announcing his presence. Broken furniture and shattered glass littered the interior, creating a chaotic scene. Navigating through the wreckage, he called out for his aunt, but only an eerie silence responded.

Whether she had been killed or abducted, it made no difference to Hindolo. His aunt possessed not a single kind bone in her body. With a resigned sigh, he turned on his heel and made his way out of the ruined apartment.

Amidst the wreckage, Hindolo's mind raced with uncertainty. Left with nowhere else to turn, his thoughts returned to the last semblance of stability in his life—the university. Reports suggested that the once lively campus lay in ruins, classrooms and lecture halls looted and set ablaze. Professors and students were now on the run, driven out by the insurgents who had occupied the grounds.

Without a clear plan, Hindolo decided to search for his former teacher and friends from the university, hoping to ensure their safety.

As he walked down the eerily quiet street, an overwhelming sense of emptiness gnawed at his core. The once bustling avenue now resembled a ghost town, a mere shadow of its former glory. Despite this, the weight on his heart remained, as he understood that no one was immune to the horrors of war. The images of his burning house, the insurgents besieging his university, and the drugged-out rebels haunted him persistently, serving as a constant reminder of the devastating reality. Uncertainty and fear had become their new normal until some semblance of peace could be restored.

Lost in his thoughts, Hindolo was startled by a familiar voice calling his name. He paused his heart racing, and turned to face the source of the sound.

"Hindolo! Wait up!" yelled Ade, one of his old friends from the university.

Ade, dressed in full army gear and holding a rifle, appeared cautious as he eyed Hindolo. "What are you doing here?" he asked warily.

Hindolo shrugged, responding, "My place got burned down. Nowhere else to go."

Understanding the feeling all too well, Ade nodded. "And the university?"

"Gone," Hindolo replied solemnly.

They stood in silence for a moment, the weight of their losses heavy on their shoulders.

"So, you joined the army?" Hindolo inquired.

Ade shifted, answering, "Yeah, given the state of things, it's not like there's much else to do, right?"

Hindolo nodded. "Who would have thought this would be our life one day?"

Ade sighed bitterly. "Yeah, apparently, college degrees can't protect us from war and chaos. Sometimes, I wonder if I made the right choice, but what else could I do? Just sit and wait for death?"

Hindolo sighed heavily, his eyes scanning the empty street. "I never imagined our country could descend into this kind of madness. It's like a nightmare we can't seem to wake up from," he lamented.

Ade nodded in agreement. "It's like everyone has lost their minds, and we're all just trying to survive. It doesn't matter who you were or what you had before. Now it's all about who has the power and who can stay alive."

Curiosity filled Hindolo as he asked, "Is the army making any progress?"

Ade nodded, his expression somber. "The state is constantly looking for more soldiers. The insurgents have been recruiting and drugging their men. Those drugged-out crazy men don't realize what they're doing. Hell, they would blow themselves up with a bomb if it meant getting their dose."

Hindolo couldn't help but sigh, knowing all too well the truth of Ade's words. He had witnessed their blind obedience firsthand.

Suddenly, Hindolo felt a surge of determination. The idea of joining the army, something he had never considered before, now seemed tempting. Ade's dry chuckle brought him back to the present.

Ade nodded, his voice tinged with amusement. "I knew this was coming. You're not the right fit, but I'm sure they can find something for you. Come on, let's go back to command and see."

Heart pounding with anticipation, Hindolo followed Ade through the dense forest. The path they took was unfamiliar, and he couldn't help but wonder where it would lead. Ade filled him in on the latest news, sharing that many of their old friends had joined the army, enticed by the promise of power and authority. They swaggered with guns, taking advantage of the weakened system to fulfill their desires.

As they ventured deeper into the forest, Hindolo couldn't shake off the feeling that he was about to step into a new world. Despite the uncertainty, he welcomed the change. Anything was better than the constant life on the run.

With each step, the chatter grew louder, guiding them toward their destination. Soon, they arrived at a makeshift command post in a grassy clearing. Men were busy with various tasks, creating an organized chaos that only those familiar with battle could comprehend.

Ade led Hindolo to a burly man dressed in a form-fitting top that emphasized his enormous biceps, cargo pants, and heavy boots.

"What's this?" the man asked, raising an eyebrow.

"A new recruit, sir," Ade responded respectfully.

The man scrutinized Hindolo, his gaze lingering for a moment before he spoke. "Well, you wouldn't last a day in the army, boy, that I can tell you. But I do have something that could prove quite useful to you," he said.

Hindolo's attention sharpened, eager to hear what the man had to offer. "Anything, sir," he replied.

The man explained, "I need you to be an informant for me. We have soldiers to fight, but we need someone who can infiltrate the rebels and pass along information. We have informants in every part of their district, but we need one in the city. You will have to be patient and stay behind the lines, passing along whatever information you can gather."

Hindolo paused, considering the dangerous task he was being offered. Joining the army had been his initial hope, but this opportunity to make a difference in a different way struck a chord within him.

With determination in his voice, Hindolo replied, "I'll do it."

Ade looked at him, surprised yet understanding. Hindolo was resolved to play his part and make a difference. This war had gone on for far too long, and if he could contribute to its end, he felt compelled to do so.

The commander nodded, calling a few more officers to brief Hindolo. Over the next three days, they delved into the intricate details of the militants' whereabouts, methods, and twisted ideology. It was like peeling away the layers of a decaying onion, exposing the grim reality hidden beneath.

Hindolo underwent rigorous training in the use of machetes and weaponry to infiltrate the enemy camp. Blending in with the violent rebels proved challenging, as they were conceited and arrogant, believing themselves unbeatable. They welcomed new recruits, assuming no one would betray them.

Playing a dangerous game of deception, Hindolo prepared meticulously for weeks until the command deemed him ready. With steely resolve, he made his way to the city, fully aware of the dangers ahead. Surprisingly, gaining access to the inner circle of insurgents was easier than expected. Hindolo joined the line of recruits, receiving a machete like a toy. It sickened him, but he played his part convincingly.

The militants had a twisted reward system. Killing a certain number of people earned narcotics, but killing high-profile targets granted them young females. Hindolo clenched his teeth, pushing through the revulsion. He couldn't show weakness and had a mission to fulfill for his country.

The militants' grip on the city tightened, leaving little room for revolt. To keep their restless men occupied, the commander planned to annex another town, demonstrating their strength. The militants eagerly prepared for battle, driven by conquest and carnage.

Within the insurgent group, Hindolo held a seemingly minor position as a busboy. Yet, he worked quietly, observing and scrutinizing everything around him. Massaging painful feet and offering water, he listened closely to the conversations, gathering crucial information. Day and night, he tirelessly cleaned up after the men, piecing together invaluable intel while remaining unnoticed.

Hindolo's time among the insurgents neared its end as the plan approached execution. With the necessary intelligence gathered, the army no longer required him to remain in the group. A surge of relief washed over him as he received the order to abandon the ship. Despising the rebels and their atrocities, he longed to be free from their grasp. Leaving the life of a spy behind, Hindolo slipped away from the insurgent base, successful in his mission.

Fleeing with a pounding heart, Hindolo knew he needed distance from the militants. He headed in the opposite direction, finding refuge in an empty village. The people huddled together in a fortified barn, filled with fear and seeking any glimmer of hope. They anxiously watched the television, yearning for news of progress or a chance at peace.

After years of struggle, the British operation Palliser, aided by a renewed UN mandate and Guinean air support, defeated the RUF, reclaiming Freetown. President Kabbah declared the Sierra Leone war over. Relief mingled with dread among the people, knowing the war's effects would linger. Rebuilding what was lost, regaining time and opportunities, seeking justice and closure—all these questions weighed heavy on their minds.

Amidst the uncertainty, Hindolo remained determined to make a difference. He refused to let the war define him and was committed to rebuilding his community. Back in college, he had encountered professors with the same goal. However, the memories of war haunted many, fearing its resurgence and the return of the RUF's chaos.

Different from those gripped by fear, Hindolo focused on his studies, determined to gain knowledge and skills to effect change.

He pursued a degree in liberal arts and humanities, eager to pick up the broken pieces left by the war.

Hindolo immersed himself in books, analyzed texts, and engaged in debates to broaden his horizons and learn as much as possible. He delved into history, politics, and society, aiming to utilize his knowledge for the betterment of his community. Graduating with his degree, he pursued law school, driven to fight for justice.

With an exceptional GPA for his bachelor's degree and an unwavering determination to fight for people's rights, Hindolo's application to law school was accepted without hesitation. He eagerly embraced this new challenge, immersing himself in the intricacies of the law, determined to learn how to leverage its power to protect the vulnerable and seek justice for the oppressed.

For three years, Hindolo delved deep into the world of law. He devoured every available book, meticulously studied case studies, and engaged in countless debates with peers and professors. He absorbed knowledge relentlessly, striving to master every aspect of the law.

Hindolo also dedicated himself to volunteering with an NGO that defended human rights. Tirelessly, he organized workshops and provided legal aid to those unable to afford it. He spent countless hours fighting for impoverished individuals who had been wronged by the system, tirelessly defending their rights and seeking justice on their behalf.

After completing his law degree, Hindolo refused to passively witness ongoing injustice in his country. Fueled by an unshakable determination to make a tangible difference, he set his sights on a new challenge—challenging the legality of a law that infringed on the fundamental right to freedom of speech.

Undeterred by the daunting task ahead, Hindolo embarked on a legal battle that eventually reached the nation's highest court—the Supreme Court. His unwavering commitment to justice, coupled with his sharp legal mind and unparalleled work ethic, made him a formidable force.

No longer the unnoticed village boy, Hindolo emerged as a fierce advocate for the oppressed—a lone voice in a country that often silenced dissent. As the legal battle raged on, his profile continued to rise. People began to take notice of his work, rallying behind him in increasing numbers. He became a symbol of hope for the silenced, shining a beacon of light in a dark and oppressive world. Despite the odds stacked against him, Hindolo refused to back down, pouring his heart and soul into the fight, driven by an unyielding conviction that justice must prevail.

Finally, the Supreme Court ruled in his favor, overturning the statute that had long silenced countless voices. It was a historic victory for all those who had suffered under repression for far too long, and Hindolo's prominence soared even higher as news of the verdict spread. He became a household name, a hero of the people, and a defender of justice. Although he knew there was much work still to be done, Hindolo was confident he was on the right path.

Hindolo's steadfast trust in the power of justice and human rights stemmed not only from his legal education but also from his faith in humanity. During his college years, he delved into the works of various humanitarians and gained awareness of the world around him. The words and acts of legendary civil rights hero Martin Luther King Jr. deeply inspired him.

Fighting for causes others dismissed became a fundamental part of Hindolo's mission. One cause that struck a chord with him was women's rights. The atrocities he had witnessed during the conflict, particularly the brutal treatment of women by soldiers and rebels, left an indelible mark on him. Horrified by the savagery, he felt compelled to assist.

The fight for women's rights held personal significance for Hindolo. He had witnessed firsthand the devastation inflicted on families, with parents abandoning their daughters out of fear. Women had become the ultimate victims of a heinous battle, trapped in a society that neglected them.

Despite skepticism and warnings to stay away from these matters, Hindolo remained resolute. He held Martin Luther King Jr.'s words close to his heart: "The time is always right to do what is right." The time was right for Hindolo. The struggle for justice and human rights was far from over, but he understood that each step brought society closer to a better future for all.

As Hindolo's passion for justice grew stronger, he embraced another cause: the fight for LGBT rights. Witnessing the increasing brutality and marginalization faced by transgender and homosexual individuals, he felt compelled to lend his voice to their cause.

Initially, Hindolo received overwhelming support and enthusiasm for his stance on LGBT rights. However, as time passed, he faced backlash from his community. Former supporters turned against him, ostracizing him for his vocal advocacy for marginalized communities.

Nevertheless, Hindolo remained steadfast. He understood the experience of marginalization and loneliness, and he was determined to make a difference with his voice. His belief remained unshakable, and he refused to be swayed by the opinions of others.

For Hindolo, the fight for justice was not merely a popularity contest. It was about upholding his beliefs and standing up for what was right, even when faced with adversity. He was prepared to sacrifice everything, including the support of those closest to him, in order to improve the lives of marginalized individuals who had endured for far too long.

The scars of war were etched deep in Hindolo's heart, serving as a constant reminder of the family and villages he had lost. He refused to stand idle as corruption and tribalism continued to tear his country apart. He was a starry-eyed young activist fueled by an unyielding fire within.

CHAPTER 7

TRAVELING THE WORLD TO EFFECT CHANGE

As Hindolo gained recognition as a human rights lawyer, he was sought after by communities from all around the world, and he decided to help others, too.

He embarked on his journey around the world. He was filled with a sense of excitement and trepidation. His first stop was South Africa, where he was greeted with curious stares and a mix of admiration and suspicion. Hindolo was ready to explore the unknown and immerse himself in a new culture. He was filled with wonder as he stepped off the plane and took in the vibrant colors and sounds of the bustling city. The people around him spoke a language he didn't understand, but he was eager to learn and communicate with them in any way he could.

The locals called him the "brown man" in Spanish, a nickname that he wore with pride but also with a sense of awareness that he was different from them. He was humbled by their generosity and moved by their resilience in the face of hardship.

As he traveled further into the continent, Hindolo was struck by the beauty of the landscape, the diversity of the people, and the richness of the culture. He spent weeks wandering through the rainforest, marveling at the exotic animals and plants that he encountered along the way. He hiked up to the peaks of the Andes, where he was humbled by the majesty of the mountains and the resilience of the people who lived in their shadows.

From South Africa, he moved on to Africa. The heat hit Hindolo like a wall as he landed in Africa. The heat scorched his skin, and the dry air seared his lungs. The terrain was harsh and cruel, with reddish-brown soil reaching as far as the eye could see. The poverty was evident, with crumbling shacks and makeshift tents along the roadways and kids begging for food.

Nelson Mandela was elected as South Africa's first black president in 1994 after apartheid was abolished. This was a historical turning point in the country's history, and it was a victory for black people everywhere. People celebrated all night, optimistic that the new administration would change their lives, especially those who had been persecuted and marginalized under the apartheid. They had suffered a lot at the hands of the all-white government that had racially segregated and oppressed them for years.

Yet, the actuality fell significantly short of many black South Africans' expectations. While there were great improvements in certain areas, such as increased political representation and a voice for black people who had previously been banned from partaking in politics and access to good education and healthcare regardless of race, the apartheid's legacy still had a significant influence on the country's economic and social frameworks.

Inequality was one of the most critical issues that South Africa was facing. The country still had one of the world's most unequal societies, with a wide disparity between the affluent and the poor. Despite the black majority having achieved considerable gains in terms of economic engagement and advancement in society, they continued to confront substantial impediments to achievement, such as limited access to resources and opportunities.

Hindolo saw the challenges the black men and women faced first-hand, and their struggle brought to light the harsh reality of capitalism.

He saw how the system favored those who were generationally wealthy and influential, while those people belonging to marginalized groups who had fewer resources were left to fend for themselves with no one to turn to and no one to ask for help.

While people had faith in Mandela and the new system he had brought following the end of the apartheid and his victory, the optimism among black South Africans slowly faded away as they saw with their eyes the progress gained was not good enough to remove the structural impediments to success and their lives remained the same. Despite the claims and promises of change, most black South Africans, including those whom he was working with, struggled to catch up with their white counterparts, who had made significant progress in the past few years.

The only people willing to help them were the legal aid clinics, various organizations, and humanitarians who could see their plight and chose to advocate for them.

Hindolo was determined to see beyond the poverty and hardship and connect with the people he had met. He wanted to learn about their culture, their struggles, and their dreams. He walked through the bustling marketplaces, the air thick with the smells of spices and sweat, and spoke to the vendors, listening to their stories with an open heart. He couldn't rely on stories narrated by government officials. All they would be concerned about was pushing their agenda. Hindolo wasn't here for politics. He was here for the people.

Hindolo then headed to Guinea. The landscape of Guinea was beautiful. Hindolo was mesmerized by it. The mountains, particularly the Fouta Djallon range, stood tall, magnificently in the country's center, with its summits exceeding a whopping 1,500 meters. The mountains were shrouded in rich green flora and different forests. They also had several lakes and waterfalls, the water clearer than the rarest crystal. The Konkouré and Gambia rivers gracefully weaved through the mountains, descending down cliffs and making stunning cascades that were visible from afar.

The raised plateau territory known as the Haute Guinée in the northeastern part was dominated by arid grasslands and savannas, where lions, hippos, and other species wandered about freely. It reminded Hindolo of his village.

The more Hindolo traveled, the more he fell in love with the country.

In the southeast region was the Guinée Forestière. It was a lush tropical rainforest with towering trees that provided a vibrant canopy over the forest floor. Many uncommon and unusual animal species existed in the region, including chimps, hippos, and forest elephants. There were also more spectacular rivers and waterfalls in the forest that Hindolo got to see such as the Chutes de la Sala and the Chutes de Kinkon.

In Guinea, Hindolo met with people who had fled their homes because of conflict, living in makeshift camps with no access to basic necessities. He was happy to listen to all they had been through and assist them in the best way that he could. They were grateful for the little he could offer them, a smile, and an ear to listen which was more than what their government had done for them.

Hindolo then headed to Uganda.

Uganda is located in East Africa and is bordered by Tanzania to the south, Kenya to the east, South Sudan to the north, the Democratic Republic of Congo to the west, and Rwanda to the southwest.

Before it had been invaded, Uganda was a place with a rich and diverse culture. It was home to multiple communities, each with its own customs, cultures, traditions, values, and beliefs.

Uganda had been home to several different kingdoms, each with its own distinct customs and practices. The Kingdom of Bunyoro-Kitara was particularly famed and well-known for its talented artists, especially the blacksmiths and potters, whose work was sought after by many countries. The country also had an abundance of agriculture. Its crops, especially millet, sorghum, and yams, were widely renowned and traded with other countries.

The Kingdom of Buganda, which ruled until the 19th century, was notable for its well-structured political framework, which comprised a centralized monarchy and an intricate socio-economic class system. The kingdom was also rich in its culture and has been known to have had a strong cultural past, including a culture of singing, dancing, and storytelling.

Uganda also quickly became a major center for trade, with merchants from all parts of the world traveling to trade various goods and purchase their ivory and gold.

Through the centuries, many other small kingdoms emerged, such as those of Ankole and Busoga, until eventually, the Europeans arrived and colonized the region, leading to a significant downturn in the country's rich history.

The East African Rift Valley, which stretches from north to south through Uganda, dominates the country's beautiful terrain, housing several different lakes, such as Lake Victoria, which is the largest lake in all of Africa.

Uganda also had the most majestic mountain ranges, especially the Rwenzori Mountains, which automatically caught Hindolo's eyes by its sheer majesty. It was often referred to as the "Mountains of the Moon" due to its glaciers and alpine plants. Aside from that, the country also had several active volcanoes and wildlife reserves that repopulated and protected endangered species.

In Uganda, during his travels, he met with young girls who had been forced to abandon their education and drop out of school. The burden of caring for their younger siblings had fallen onto their shoulders. They were eager to learn, but the school fees were beyond their families' means. They yearned for the opportunity to escape the cycle of poverty but remained trapped with no hope of liberty.

Hindolo felt his heartache as he saw the contrast between the natural beauty of the land and the harshness of the lives that people led. The stunning savannahs and majestic wildlife were a sharp juxtaposition to the desperation and struggle that the locals faced every day.

As Hindolo journeyed further into Africa, the landscapes changed but the hardships persisted. However, he was taken aback by the level of intolerance to dissent and criticism of the government in some countries. He witnessed firsthand how those who dared to raise their voice against those in power were silenced, detained, and subjected to unspeakable torture. It was as if the governments saw any form of opposition as a threat to their power, and they were willing to do whatever it took to quell it.

As he met with activists who had been persecuted, assaulted, and imprisoned for speaking up, Hindolo felt an overwhelming feeling of dread sink in his chest. He heard stories about journalists who went missing after writing critical pieces and about students who were shot for peacefully demonstrating. The more he studied, the more he realized how bad things were and how hard this would be.

Despite the danger, Hindolo realized he couldn't keep quiet. He utilized his voice to bring attention to the injustices going on, publishing articles and delivering lectures wherever he could. He connected with other activists and politicians striving for change, and together, they tried to shed light on the continent's darkness.

But it was not without consequences.

He was startled by an echo of boots coming up the stairs. The thick soles clashed against the weak floor, the sound reverberating through the paper-thin walls. He sprang out of bed, his heart beating so hard against his chest he felt like it would burst. He had been warned that the officials might try to abduct him, angered by the work he was doing. In such an event another friend of his who was a prominent lawyer had asked Hindolo to call him. Hindolo scrambled towards the phone, but before he could get his hands on it, the door to his room slammed open, and multiple men in uniforms swarmed in, capturing him forcibly and hauling him outside.

They had cut the power of the building, and Hindolo could not make out their faces. As they brought him outside the hotel, he could see the menace in their eyes, highlighted by the moon's glow shining brightly in the dark night. He tried to protest, to ask what was going on, but they refused to answer him and instead shoved him into a waiting car. They dragged him inside, blindfolding him with a brown sack, obscuring his vision, and shoving something in his mouth to drown out his shouts.

He had no idea where he was being taken.

The car sped off into the night. He could feel two men sitting on either side with rifles between their legs. They had tied his hands behind his back. All he could do was listen, for his screams fell on deaf ears. As the car raced ahead, making rough turns, his body would fling around, and his captors would shove him back into place.

As the Colombian officials escorted Hindolo into the dark, damp halls of the prison center, his heart pounded rapidly. His mind was racing with the worst-case scenarios. He did not know what was happening. Would they torture him? Would they kill him? Surely, they wouldn't kill him.

He knew whatever claims they had against him would be unfounded and baseless. No court in their right mind would prosecute him, but he also knew that in this part of the world, the accusations alone were enough to cost someone their life.

He was thrown into a tiny cell. He screamed and shouted, demanding a lawyer, begging to know what he did wrong, but they refused to answer. He stayed in the cell for what felt like days. There were no lights no windows, save for a tiny bulb in the middle of the room. Hindolo had no way of knowing what time it was.

Moments later, the cell door swung open, and three men strode in, their presence sending a chill down Hindolo's spine.

His fury was palpable as he demanded answers, "What the hell is the meaning of this? What have I done?"

One of the men, dressed in a black shirt and cargo pants, flashed a menacing smile, "You are being investigated for treason. We suspect you of being a CIA agent - a conniving spy sent to infiltrate our country and stir trouble. I will ask you some questions, and you must answer them truthfully. We can do this the easy way or the hard way. The choice is yours."

.The other two men, dressed in military attire, stood stoically in the background. All the color drained from Hindolo's face. What did they mean? How could he be a spy for the CIA? And most importantly, why were these three men here?

"Now, tell us," said the man, taking a menacing step forward toward Hindolo, "Who sent you here?"

Hindolo's heart raced as he looked up at the towering figure in front of him. He knew he had to choose his words carefully, "No one! I am an activist fighting for the people! I have come here with the sole intention of helping them." he frantically explained.

A look of disappointment crossed the man's face as he let out a menacing sigh, "I had hoped you would cooperate, but it seems we'll have to do this the hard way."

Before Hindolo could react, the man's fist collided with his face, sending him reeling backward. Pain exploded through his skull as he fell to the ground, dazed and disoriented. The other two men remained silent, their eyes staring ahead blankly. Hindolo attempted to get up, but the two men suddenly walked forward and pinned him down.

He was accused of being a spy and a traitor to the country. He objected, but his protests were ignored. Instead, they continued to use physical and psychological abuse to break him down piece by piece.

For days, he was subjected to relentless interrogation and harassment. He denied the charges vehemently, but the authorities seemed unwilling to believe him. Instead, they subjected him to cruel and inhumane treatment, depriving him of food and sleep and subjecting him to physical and psychological torture.

Every day, it was the same thing, over and over again.

Hindolo groaned in agony as he struggled to sit upright. Blood oozed from his swollen lip, and his head throbbed relentlessly, but he refused to back down. He knew that he had to prove his innocence, no matter what it took.

"Who are you working for?!" the interrogator barked, his eyes blazing with fury.

"I already told you, I'm fighting for the people! I'm not a spy!" Hindolo protested, his voice weak and hoarse.

The interrogator leaned in close, his breath hot on Hindolo's face. "Don't lie to me! We have evidence that proves you're a CIA agent!"

Hindolo's heart sank at the mention of the CIA. He knew that if they truly believed he was working for the Americans, his fate was sealed.

But he refused to give up. With every shred of intelligence he could muster, he tried to outsmart his interrogators.

"I demand to see this evidence you have against me!" he said, his voice gaining strength. "If you have nothing to hide, then you have nothing to fear."

The interrogator sneered at him, "You're in no position to demand anything, traitor!"

But Hindolo refused to be intimidated. He continued to argue, to question, to demand proof of his alleged wrongdoing.

Despite the agony he was experiencing, Hindolo refused to give up. He knew that he was innocent, and he was determined to clear his name. He used every ounce of his strength and intelligence to outsmart his interrogators, to prove to them that he was not the enemy they believed him to be. He would use logic to defy any evidence or fact they produced to link him to the claim of being a spy.

Finally, after weeks of being detained in harsh conditions, the authorities realized that they had made a mistake. They had no evidence to back up their claims, and they knew that holding Hindolo any longer would only cause more problems. They decided to release him.

Hindolo shaded his eyes from the unexpected brightness of the outside world when the hefty metal door of the cell cracked open. They were finally here to tell him that he was free to go. His months in captivity had left him thin and feeble, but as he stepped out into the sunlight, he felt a glimmer of hope. He couldn't believe it was all over. He'd been confined for weeks in a location where time appeared to stop still, and every moment seemed to last forever. But he was now finally free.

He felt a lot of different things as he stepped out of the prison gates: relief, thankfulness, and wrath. He couldn't believe he'd been jailed for so long with no proof to back up the claims leveled against him. Even though he had escaped, he knew the experience left a permanent mark on him that he would carry for the rest of his life.

As he left Colombia behind, Hindolo felt a renewed sense of purpose. He knew that there were countless others out there who were being falsely accused, harassed, and detained for daring to speak out against injustice. He knew that his experience could have ended very differently, and he was determined to use his voice to fight for those who could not fight for themselves.

Hindolo continued his journey, driven by a sense of purpose and a deep commitment to the ideals of justice and good governance. He became a tireless advocate for political reform, using his voice and his platform to push for change at the international level. He met with world leaders, diplomats, and activists, calling for greater transparency, accountability, and respect for human rights.

He had the honor of meeting some of the most remarkable leaders of all time while traveling the world, speaking out against political intimidation, and campaigning for human rights diligently.

He had the honor of meeting the United Nations Secretary-General. He was an influential figure in the world of politics and a great advocate of peace, working relentlessly to support the oppressed. The encounter with him was both humbling and uplifting for Hindolo, as he could see that his efforts were part of a worldwide movement and he was on the right path.

Another incredible experience was meeting Nelson Mandela, his personal hero and the respected politician and leader who had been imprisoned for 27 years for his opposition to apartheid in South Africa. Meeting Mandela was truly humbling for Hindolo. He was astounded by the man's intelligence, kindness, and humility. Mandela's unshakable dedication to justice and equality for his people motivated him to step up his own efforts in the fight for human rights.

He also had the opportunity to meet Kim Campbell, the former Prime Minister of Canada and ardent supporter of women's rights. Their talk was illuminating, as she discussed her own experiences with overcoming gender prejudice in politics and public life. He was captivated by her tenacity and steadfast dedication to making the world a more just and equal place. It was truly inspiring.

As he traveled the world, Hindolo discovered that his true calling was not just to see the world and experience its wonders but to use his experiences to make a difference.

He then went to Kenya. When he landed and began traveling, he was struck by the stark divide that existed between the natural beauty of the area and the horrible history of the British invaders who had committed violence and cruelty against the people of the Maoedin society.

The Mao Mao uprising was an armed struggle between the British colonial government and a group of Kenyan rebels fighting for independence from British rule known as the Mao Mao.

Once the rebellion had been quashed, the British colonial authorities used collective punishment to punish entire villages for the actions of a few rebels. As a result, the British used ruthless and violent measures such as torture, rape, and extrajudicial murders. The rebels' wives and children were captured to torture the rebels. They were forced to watch their families suffer as retribution.

Many Mao Mao insurgents were captured and imprisoned in concentration camps and subjected to brutal punishment. They were forced to provide hard labor, deprived of food to the point of starvation, and beaten black and blue, physically assaulted whenever the soldiers felt like it. British authorities also used forced relocation as a means of disrupting the rebels' support network and denying them food and shelter, making them homeless and reducing them to the status of beggars.

He met with the survivors and heard terrible stories of the torture and brutality that had barbarically been perpetrated on members of the freedom fighter organization. He was shocked and devastated to hear about the atrocities that had been committed against them and the lack of attention the survivors had received. Nearly 11000 Kenyans were killed, he couldn't understand why more people weren't angry.

He was astounded by the richness of natural beauty, from the magnificent wildlife to the different landscapes. Kenya was stunning. However, even the beauty of the land couldn't distract him from the sharp difference that existed between the prosperous

Indian community, which was thriving while the black community suffered. The discrepancy in income and standard of living was evident, and it served as a reminder that colonialism and tyranny remained throughout the country despite having ended officially.

As he continued to explore Kenya, he wondered why such inequity and injustice remained and his quest to help the people further strengthened.

By now, he had gained recognition worldwide, and people were approaching him to partake in summits and conferences in struggling countries to advocate for their rights.

He was invited to the Philippines for a conference which was taking place in Manila. Hindolo was captivated by the natural splendor of the place. The Philippines, with its many islands and beautiful beaches, provided a peaceful respite from the city's rush and bustle.

He learnt about the nations' fights for independence and the impact of colonialism as he dug more into their histories before attending the conference. He was particularly touched by stories of the Filipino resistance and fortitude under Spanish, American, and Japanese colonial authority, as well as the war for independence led by men like Jose Rizal and Andres Bonifacio. These people had suffered at the hands of their colonizers and were left scarred as a result, with no one to protect or advocate for them.

The global summit in Manila brought together international leaders and activists to discuss the significance of human rights and the democratic government. He was thrilled to be a part of it. His main goal was to be the voice of the oppressed and an advocate for development not just in his own region but for those suffering around the world. He also observed the Philippines' authoritarian police state, where President Duterte's aggressive drug campaign resulted in thousands of extrajudicial deaths and severe human rights violations. The government's attack on the press and political opposition was particularly concerning, indicating a potentially serious deterioration of democracy and civil freedoms, which Hindolo spoke up against at the conference.

He was on a mission to spread the message of equity and social justice throughout the world. The more conferences he attended the more he was invited to. Soon enough, invitations started pouring in from all over the world. These would be sponsored, and it would be a great opportunity to speak for his people. He gathered his things and set out on his trip to Europe.

He marveled at the diversity of cultural groups and the beauty of the places he saw as he traveled from nation to country. In England, he was captivated by London's rich history and architecture, while in Belgium, he relished in the country's world-famous chocolates and waffles. He was captivated by the spectacular natural splendor of the fjords and mountains in Norway and Sweden. In France, he enjoyed the great cuisine and culture.

Despite the beauty and splendor, he couldn't help but observe the vast disparity that existed in living conditions between the West and the rest of the world. He questioned why some regions had so much prosperity while others struggled to fulfill even their most basic necessities.

To better understand this, in his free time, he dug further into the role global organizations like the World Bank and the IMF play in influencing the global economy. The more he studied, the more he became acquainted with the intricate mechanisms of wealth distribution as well as the power structures that perpetuated and exacerbated inequality.

He slowly began to conclude that these organizations' policies have often exacerbated inequality and caused harm to the most vulnerable populations rather than looking out for them. He realized that the institutions prioritized the interests of wealthier countries and companies over the needs of poorer countries, resulting in rising debt and economic insecurity.

Somalia was struggling quite badly, and despite plenty of activists telling him not to go, he decided to go there. He felt a knot grow in his stomach tightening as he got off the plane after it landed. The air was thick with tension, and explosions could be heard in the distance. The streets were strangely quiet, as expected. After all, the country was in a state of war, and the few individuals he saw moved quickly as if scared they'd be dead any minute. His heart was racing as he walked to his hotel, each step seeming like a bet on his life.

The hotel was a fortress with thick walls and armed guards. He couldn't help but feel imprisoned within them, with the continual fear of danger lurking right outside the walls. He found himself continuously looking over his shoulder, searching for any signs of danger.

But it wasn't simply the physical risk that worried him. The feeling of gloom and hopelessness that lingered over the country like a black cloud was nearly stifling. He encountered people who had lost everything as a result of the region's violence and disarray, their lives destroyed by the incessant upheaval, their families torn apart. Their helplessness broke Hindolo's heart.

While in Somalia, his hotel was attacked twice.

When his hotel was attacked for the first time, he had just come back from a long day of meeting survivors of the war and getting their stories. He had traveled all day and was completely exhausted to the point where he came to his room and dropped on the bed. Just then, a powerful blast rattled the structure, causing the ground to shake violently. As he rushed for refuge under the bed, he could hear glass shattering and rubble falling all around him. The air was heavy with dust and smoke, making breathing difficult. His heart nearly exploded when he realized he was in the midst of a terrorist strike.

The security guards got him out before the building swallowed him whole. They immediately relocated him to another hotel. However, that, unfortunately, was not the end of it.

The second attack occurred just days later as he was preparing to leave the country. He'd barely finished packing his things when he heard a huge explosion. He dashed to the window, where he noticed smoke billowing from the next street. He knew he had to move quickly if he was going to make it out alive for the next target would be the hotel. He gathered his things and rushed down the steps, evading shattered glass and rubble, protecting himself as best as he could. He could hear gunshots in the distance and sirens approaching. He ran as fast as he could out of the building and into the mayhem outside, escaping death by a hair.

After the attack he immediately headed to the airport and boarded his flight to Somaliland, pondering over how fleeting life can be.

The atmosphere, however, changed dramatically when he journeyed from Somalia to Somaliland. The sounds of gunshots and explosions were replaced by the sounds of chuckling and chatter from the locals. The streets were alive with bustle, and the marketplaces were brimming with vibrant products. The people he encountered were kind and ready to share stories about their culture and history.

He wanted to understand why Somalia struggled in such a way when Somaliland was thriving. He observed the people and talked to the elders. This was when he came to realize why Somaliland was so different from its turbulent neighbor. Despite the fact that it was not recognized as a sovereign state by the international world, Somaliland had managed to develop a functional government and society. People had banded together to re-establish their country from the rubble of civil war, and they were driven to succeed.

He was captivated by the people of Somaliland's tenacity and persistence. Against all odds, through several hurdles and barriers, they were able to establish a secure and peaceful community.

Hindolo was also invited to Jordan's capital, Amman. During his journey, he was able to explore the West Bank properly, seeing the sights and visiting important places. He even got permission from the state to travel to Israel and explore the city.

Despite how close the two countries were, Hindolo was taken aback by the mutual hostility that existed between them and how much the two countries hated each other. They had a complicated history which contributed to it.

The land had been partitioned and distributed between Israelis and Palestinians. The battle over the land began in the late 1800s when Zionist Jews began to immigrate to Palestine to escape Germany in search of a new homeland. Back then, it was under the control of the Ottoman Empire, which operated as one country.

Following World War I, the British acquired control of Palestine, and the United Nations agreed in 1947 to divide the area into two nations, one for Jews and one for Arabs. However, the Arab states rejected this choice, resulting in a war between Israel and its neighbors in 1948.

The irony of religious prophets like Abraham, Moses, and Jesus being born in the same place, yet both divided by manmade lines due to the enmity sown among their followers, did not go unnoticed by Hindolo. He was, however, intrigued by the rich history of the region, and he wanted to learn more about the three main religions that divided them.

In Jerusalem, he visited the Western Wall and the Church of the Holy Sepulchre, as well as the Dome of the Rock in the Al-Aqsa Mosque. The majesty and historical importance of these locations astounded him as he marvelled at the artifacts and architecture of them understanding why people from all over the world had been coming to these sites for ages.

He came to understand that the world was full of challenges but also full of opportunities for those who were willing to fight for what they believed in. And so, he continued on his journey, driven by a sense of purpose and a determination to make the world a better place.

CHAPTER 8

POLITICS OR HUMANITY

As Hindolo stood at the crossroads of his dreams, his mind filled with uncertainty and indecision. The weight of his choices pressed upon him, mirroring the complex web of emotions that entangled his heart. The voices within his troubled mind engaged in a spirited dialogue, each representing a facet of his soul, each vying for dominance.

"At the peak of your career, Hindolo," one voice whispered, "the world beckons you towards the realm of national politics. Imagine the impact you could have on a grand scale, shaping policies, influencing legislation, and defending the defenseless from within the system. Your unwavering commitment to human rights would be magnified, reaching far beyond the confines of your current sphere."

Another voice, tinged with caution and skepticism, chimed in, "But at what cost? Look around you, Hindolo. See how politics has changed many, eroding their once noble intentions. Politicians are known for their lies, empty promises, and their willingness to compromise on matters of principle. How can you, a sincere and honest rights-based activist, cross the line into a realm stained with such corruption?"

The weight of history bore down on Hindolo's thoughts as the specters of Sierra Leone's past emerged from the recesses of his mind. The tragic fates of political opponents, falsely accused of treason and executed by hanging, haunted his conscience. The inhumanity of the death penalty, exploited and abused in third-world countries, gnawed at his soul. Freedom of expression, a pillar of democracy, was under attack, with values being cast aside in favor of the prevailing norm.

Dawad Hindolo was reminded of some piteous murders carried out in his country by sitting presidents against opposition members, even opposition within the same political party. They became targets of unleashing murderous charges that witnessed death only by hanging. Abuse of power galore in the crudest sense of the word. Just after arriving in the city, Hindolo was fascinated by the history of his country. He learned about the beautiful tribes and people and the languages, which he learned through an old man, Pa Senesie, whom he would spend a long time with.

The old man explained to him about the different districts and the battle between the British colonial masters and Bai Bureh. Pa Senesie explained how a local chief could easily challenge the almighty colonial regime. During his sojourn, he learned about the death through the hanging of prominent politicians in the earliest part of history: Dr. Mohamed Sheku Fornah and Brigadier Lansana. These lowest displays of inhumanity etched a mark in his mind, and now that he is about to jump into politics, he is worried that the fate of these two may just befall him.

Pa Senesie had pitched his interests that Dr. Fornah was a reformist medical doctor who was invited by the Prime Minister, Siaka Stevens, from the United Kingdom to join his government. As they ruled the country, Dr. Fornah realized that Siaka Stevens was a dictator and had little interest in moving the country forward. Pa Senesie went on, saying that Dr. Fornah, a highly educated individual, wrote a public letter to the half-baked trade unionist Prime Minister, which became his footpath to the gallows. He went on musing that maybe the expressions were too sophisticated for the Prime Minister and got him humiliated.

Dr. Fornah wrote:

"Dear Prime Minister,

In 1967, Sir Albert brought this country to the brink of political and economic disaster. All right-thinking people realized that Sir Albert's insatiable desire for power and wealth spilled chaos and complete disruption of our social fabric.

What he wanted to impose upon this nation was a one-man dictatorship shrouded by a fraudulent Republican Constitution that concentrated all powers in the hands of a single person.

In view of this menace to personal freedom and economic stability, I accepted the call to service and left what you knew to be a very lucrative medical practice to join the fight against that political monster.

As the leader of the Opposition, you spearheaded the fight against this menace. All of us who followed you accepted your profession of a deep attachment to the tenets of democracy and the rule of law. As you also know, when the Military unwarrantedly usurped the machinery of Government and imposed a Military dictatorship on the people of this country, you found me more than willing to risk life and limb to restore Parliamentary democracy to this beloved country.

It is nearly three years now since we assumed the reins of Government. Over this period, I have had the opportunity to work closely with you. Many mistakes have been made during this period, but until lately, I had assumed that these were mistakes of the head and not of the heart. I now know that I was wrong.

You have revealed an uncanny dexterity at manipulating weak and untutored minds both at Cabinet and Party levels. Your conduct of state affairs is in line with your Trade Union experience, a mixture of trite jokes, cajolery, and even violence.

The introduction or the continuation of cold and calculated violence into the politics of this country poses a major threat to the social and political unity of this country. You are fully aware of the events I refer to.

The shootings at the Freedom Press, resulting in the tragic death of an innocent child, have never been satisfactorily investigated to the thinking public's satisfaction. The wanton destruction of life and property at Ginger Hall during the Freetown City Council elections was both unprovoked and unnecessary. As usual, it was the innocent, especially innocent children, who suffered. Once again, the nation has been left waiting in vain for an explanation.

More recently, equally serious and despicable events occurred at Port Loko. I squarely place the blame for these heinous crimes at your doorstep. I know, as you do that you were the mastermind behind them. I have spared no effort, both inside and outside of Parliament, to condemn the use of force and violence as a means to attain power. This fundamental disagreement between us is a point of contention.

A Constitutional Review Commission has been established and is currently at work. However, during the interim, with you in charge, a so-called National Executive of the APC passed a resolution in favor of an Executive Presidency, consolidating all state powers in the hands of one individual. It is no secret that you have always harbored ambitions of becoming such a President.

You initiated this course of action during the meeting I refer to. I distinctly recall the ongoing struggle between you and the Protocol Officers regarding the playing of the National Anthem for you. His Excellency the Governor-General has, on more than one occasion, been subjected to significant embarrassment regarding the use of the National Anthem when both of you attend the same functions.

While this display of vanity may seem insignificant to some, as someone with a trained medical mind, I see them as manifestations of a megalomaniac syndrome. They represent just the tip of the iceberg, with a sea of personal shyness lurking beneath. Coupled with an insatiable thirst for power, this can only spell disaster for our country. I now realize that you are willing to employ any means necessary to achieve this goal.

Upon my return from the USA and Britain, I made my views on this matter abundantly clear to you. I explicitly stated that if you insist on pursuing this course of action and introduce a bill in Parliament advocating for an Executive Presidency, I will have no choice but to oppose it with all the means at my disposal. As is typical with you, your response to a significant issue was not an enlightened discussion but rather an overused joke.

I have observed that our differences regarding financial policy have deepened over time. The Cabinet reshuffle in May of this year was triggered by the refusal of the then Minister of Works and myself to approve the expenditure of six million leones of taxpayers' money on constructing a rock-filled road that was less than two miles long. As I indicated in a letter to you, this project had no economic merit and did not address the purported traffic congestion issue used to justify its conception.

The use of supplier credit and pre-financing in the public finance of an underdeveloped country has limited scope. Its misuse and abuse led to the near bankruptcy of the nation between 1966 and 1968. You yourself have condemned this practice in your radio broadcasts to the nation.

I have always maintained that pre-financing should be limited to a small number of projects that contribute to the social well-being of our people. Examples of such projects include providing clean, piped water and expanding our radio services.

In my discussions with World Bank and IMF officials, I have expressed our general agreement with them to avoid unnecessary pre-financing. You yourself have reassured the World Bank along similar lines.

However, I have recently noticed that rejected pre-financing schemes are being resurrected and brought to the Cabinet without my Ministry being given sufficient opportunity to comment on them. During my brief absence on a duty tour, you signed a contract against the advice of my officials to pre-finance the purchase of armored vehicles costing nearly Le700,000.

At a time when soldiers are poorly housed, I question whether your priorities as Minister of Defence are in the right order. The military was never given the opportunity to express their opinion on this scheme.

Furthermore, you have presented an elaborate scheme for police and army communication equipment. If accepted, it will cost the nation Le2.8 million. What I find perplexing is that the communication officers of both the army and police have estimated their requirements at Le300,000, roughly one-tenth of the cost of your Pye advisers.

The curious combination of your relentless pursuit of an Executive Presidency and the acquisition of armored vehicles to be manned by specially selected troops primarily loyal to you suggests that you intend to impose your will on the people of this country.

You have consistently compared yourself to Sekou Toure, Kaunda, or Nyerere. However, Sierra Leone is a distinct country, Mr. Prime Minister, and to be frank, you are not a Kaunda or a Nyerere.

Government policies in other areas are equally perplexing. We are all aware of the repeated use of troops and police to evict undesirable elements from Kono. However, as the country knows, this has become a cyclical exercise. You publicly drive them away with great fanfare but quietly allow them to return for reasons that only you can understand.

There is a lack of coherence in government policy, a lack of definitive coordination of policies, and a lack of firmness in executing policies.

Your Kono exercise cost the nation an additional Le600,000 for the army and police in 1969. The bill for the latest exercise is still pending. Meanwhile, most of the Lebanese individuals expelled through this exercise have now been permitted by you to return.

If our limited resources are not used to create gainful employment for our people, then, as I have repeatedly emphasized to you, we are witnessing a form of highway robbery. These problems cannot be solved by armored vehicles or overly expensive communication equipment.

I have expressed my views at length to ensure that the nation understands. I am painfully aware that my resistance to these schemes has not endeared me to you. However, I owe it to myself and my country to fulfill my duty as I see fit. I had intended to attend the crucial meetings of Commonwealth Finance Ministers in Nicosia and the World Bank and IMF in Copenhagen. Yet, I have received reliable information that you intend to revoke my accreditation to these conferences upon your return.

The honor of this country is of paramount importance to me, and I will not allow you to further ridicule our nation, as you have done in the recent past.

Finally, I caution the nations of the world that if their citizens allow you to engage in a pre-financing spree during the final days of your regime, this nation reserves the right to repudiate these debts in the future.

Given these tensions and our fundamental differences in principles and policies, I realize that my usefulness in this government under your leadership has reached a nadir. I cannot sacrifice principles for a position, but as I always say, let history be my judge. Therefore, I hereby tender my humble resignation, effective immediately."

Yours sincerely,

(Signed) Dr. M.S. Forna

Pa Senesie, with a vivid expression, bowed his head for a moment and continued, "A sad chapter was written in the political history of Sierra Leone when Dr. Mohamed Sorie Forna, one of Sierra Leone's most brilliant politicians, along with Lieutenant Habib Lansana Kamara, Ibrahim Bash Taqi, Brigadier David Lansana, and Paramount Chief Bai Makarie N'silk, were executed at the notorious Pademba Road Prison in Freetown."

Biting his lower lip and holding Hindolo by the shoulder, he went on, "The pitiful sight outside Pademba Prison, where the men who had been hanged were displayed for about an hour to convey a strong message, sent shockwaves throughout Sierra Leone and even beyond."

"Was there a fair trial or any trial at all?" Hindolo asked. With wide eyes and a trembling voice, Pa Senesie educated the inquisitive Hindolo, saying, "The marathon trial of Mohamed Sorie Forna and fourteen others captivated the nation. The eventual execution of Dr. Forna and Ibrahim Taqi, a former Information Minister, two distinguished Sierra Leoneans who played a pivotal role in the All People's Congress' election victory in 1967, was not only a source of grief for their families and loved ones but also a tremendous loss for the entire nation. It was a national calamity that had immediate and far-reaching repercussions. The executions alienated the support of many northerners, especially in Tonkolili, an APC political stronghold during the 1967 General Election, and fueled the flames of popular discontent against Siaka Stevens."

Pa Senesie then explained that this was not the only incident. "Stevens' popularity as the leader of the All People's Congress and Prime Minister of Sierra Leone started to decline in 1971 when his government executed Brigadier John Bangura, the man who had handed over power to him in 1968, along with Jawara and

Kolugbonda. These three individuals were the first high-ranking officers to be executed in Sierra Leone after independence in 1961. Brigadier John Bangura wept at the eleventh hour of his execution, finding it hard to believe that Siaka Stevens, the man to whom he had handed power, rejected his plea for mercy. It was also revealed that Bangura never faced the gallows; he was beaten to death when he refused to walk to the gallows. What a terrible way to end for a decorated officer held in high esteem!"

Hindolo now recollects these stories, and the prospect of a recurrence of history sends a chill down his spine. He takes solace in the development of human rights in his native country, the global condemnation of brutality and suppression, and the emergence of social media, which exposes such despicable acts of eliminating political opponents.

But even in the midst of this darkness, Hindolo's thoughts turned towards his family. His wife and two children depended on him for protection and sustenance. Serving humanity had been his guiding light, but the harsh reality was that it did not always put food on the table. Donor support, once plentiful, had dwindled, leaving many human rights and democratic activists grappling with uncertainty.

The turmoil within Hindolo intensified as he grappled with the impossible choices before him. Saving the world or ensuring the well-being of his loved ones seemed to be an irreconcilable conflict. Yet, deep within his soul, a flicker of clarity emerged—a realization that perhaps he could forge a path that harmonized both.

The dream landscape shifted, and Hindolo found himself transported to distant lands, his footsteps tracing a path across the

112

globe. His journey began in South Africa, a nation marked by its tumultuous past and its resilient spirit. The people there greeted him with curiosity and a mixture of admiration and suspicion. The vibrancy of the culture, the vibrant colors, and the bustling city awakened his senses. Despite the language barrier, Hindolo was eager to learn and communicate, immersing himself in this new world.

The people bestowed upon him the moniker of the "brown man" in Spanish, a nickname he wore with pride, a testament to his differences and his unity with them. Hindolo was humbled by their generosity and moved by their unyielding spirit in the face of hardship. It was in these moments of connection that he found renewed inspiration, a reminder of the resilience and strength of humanity.

As Hindolo journeyed further into Africa, the land embraced him with its searing heat and rugged terrain. The reddish-brown soil stretched endlessly, a testament to the harshness of life in this region. Poverty reared its head, with crumbling shacks and makeshift tents dotting the landscape, and the specter of hunger haunted the youngest and most vulnerable. Yet, amidst the struggles, Hindolo witnessed the indomitable spirit of the African people, their resilience shining through the darkest of circumstances.

His thoughts turned back to South Africa, where the triumph over apartheid had been a landmark victory. Nelson Mandela's election as the nation's first black president in 1994 ignited a flame of hope in the hearts of black people everywhere. The abolishment of racial segregation was a turning point, a testament to the power of unity and collective action. However, the reality fell short of expectations. The legacy of apartheid continued to cast a long shadow, affecting the country's economic and social fabric.

Hindolo recognized the complexities of politics and its ability to uplift and disappoint in equal measure. He understood that navigating this treacherous landscape would require unwavering principles, a steadfast commitment to justice, and an unyielding dedication to the values that had guided him thus far.

In this realization, Hindolo made his decision. He would embrace the realm of national politics not as a surrender to corruption but as a vessel for change. He would become the voice of integrity, striving to dismantle the corrupt systems that had stained the political arena. With his unwavering commitment to human rights and justice, he would be a beacon of hope, fighting for the vulnerable, defending the defenseless, and protecting the values he held dear.

The crossroads no longer posed a dilemma but rather an opportunity for Hindolo to merge politics and humanity, to bridge the gap between serving the world and safeguarding his family. He would chart a course that balanced both, drawing strength from his experiences across the globe and forging a path that would honor his purpose.

With a resolute heart, Hindolo stepped forward, his footprints etching a new chapter in his life. The crossroads of his dreams now represented not only a junction of choices but also a convergence of his principles, his family, and his unwavering dedication to a just and compassionate world.

As Hindolo embarked on his political journey, he knew he couldn't do it alone. He reached out to fellow activists, human rights organizations, and individuals who shared his vision.

Together, they formed a coalition, a united front against corruption and injustice. Their shared goal was to transform the political landscape, to inject integrity and compassion into the heart of governance.

Hindolo's decision to enter politics didn't come without sacrifices. He knew that his new path would demand immense dedication and perseverance. The campaign trail was grueling, filled with long hours, sleepless nights, and countless meetings with constituents. But Hindolo remained steadfast, fueled by his unwavering belief in the power of change.

As he crisscrossed the country, connecting with people from all walks of life, Hindolo realized the true power of his position. He had a platform—a platform that could amplify the voices of the marginalized, shine a light on the issues that had long been ignored, and bring about tangible change. He spoke passionately about human rights, about justice, and about the need for a government that truly served its people.

Slowly but steadily, Hindolo's message began to resonate with the electorate. People saw in him a leader who wasn't afraid to challenge the status quo, a leader who would fight tooth and nail for what was right. They saw a man who had walked in their shoes, who had witnessed firsthand the injustices that plagued their lives.

The day of the election arrived, and Hindolo stood before a sea of faces filled with hope and anticipation. As the votes were tallied, it became evident that the people had spoken. Hindolo emerged victorious, elected as a representative of the people, a voice for the voiceless.

But Hindolo knew that his work had only just begun. The real challenge lay ahead in the halls of power, where corruption lurked and compromise whispered its seductive promises. He would face temptations, challenges, and setbacks, but he remained resolute in his mission.

In his new role, Hindolo wasted no time in introducing legislation that prioritized human rights, social justice, and transparency. He worked tirelessly to strengthen the institutions of democracy, to empower the marginalized, and to hold those in power accountable. He surrounded himself with like-minded individuals and experts in their fields who shared his vision for a better society.

Change didn't come easily, and Hindolo faced opposition at every turn. But he drew strength from the stories of those he had met along his journey—the resilient woman fighting for clean water in a remote village, the young boy dreaming of an education that seemed out of reach, the family torn apart by violence. Their struggles fueled his determination, reminding him of the urgency of his mission.

As Hindolo's tenure in politics continued, he became a beacon of hope for the nation. His unwavering commitment to justice and integrity inspired others to join the fight to believe that change was possible. The corrupt were exposed, the injustices dismantled, and step by step, the nation began to heal.

Hindolo's decision to enter politics was not without sacrifices, but the rewards were immeasurable. He saw firsthand the impact he could have on a grand scale, the lives he could touch and transform. And amidst it all, he never lost sight of his family—the driving force behind his journey.

His wife and children stood by his side, providing unwavering support and love. They understood the sacrifices he made, the long hours and sleepless nights. Together, they navigated the challenges of public life, finding solace in the knowledge that they were working towards a better future for all.

As Hindolo looked back on his journey, he marveled at how far he had come. The crossroads of his dreams had led him down a path he had never imagined, a path that merged politics and humanity, a path that had brought him closer to realizing his purpose.

And so, Hindolo's story continues—a story of resilience, of unwavering principles, and of the power of one individual to make a difference. As he forges ahead, he carries with him the voices of the marginalized, the hopes of the oppressed, and the belief that, together, we can build a more just and compassionate world.

CHAPTER 9
PEOPLE AND CULTURE

The sun rose on the horizon, casting a warm golden glow over the bustling streets of Freetown, Sierra Leone. Hindolo emerged from his humble abode, taking in the sights and sounds that greeted him each day. In these streets, he had endured hardships, fought for justice, and now stood at the precipice of a new phase in his life.

As the morning breeze caressed his face, Hindolo's thoughts turned to the places he had been, the cultures he had encountered, and the stories that had shaped him. He embarked on a voyage through his memories, reliving the diverse landscapes and vibrant tapestries of humanity he had witnessed during his travels.

Indonesia was his first stop on this journey of reminiscence. The country beckoned with its captivating blend of history, culture, and natural beauty. Stepping foot into this vibrant land, Hindolo found himself immersed in a sensory tapestry that awakened his senses and etched indelible memories in his mind.

The bustling markets of Indonesia were a testament to the vibrant energy that permeated the country. Hindolo could vividly recall the sounds of vendors calling out their wares, the chatter of locals engaged in animated conversations, and the rhythmic melodies of traditional music that echoed through the air. The markets were a kaleidoscope of colors, with vibrant fruits, spices, and textiles adorning the stalls, creating a feast for the eyes.

Amidst the bustling markets, Hindolo encountered an intoxicating symphony of scents as the aromas of exotic spices wafted through the air. The fragrant notes of cloves, nutmeg, and cinnamon mingled harmoniously, enveloping him in a sensory embrace. Each inhalation brought forth a flood of memories, reminding him of the rich culinary traditions and diverse flavors that defined Indonesian cuisine.

But it wasn't just the grand landscapes and historical narratives that captivated Hindolo. The warmth and kindness of the Indonesian people left an indelible mark on his soul. They welcomed him with open hearts and genuine smiles, displaying a hospitality that resonated deeply within him. Whether he was strolling through the bustling streets of Jakarta or venturing into the tranquil villages of Bali, Hindolo was met with genuine curiosity, openness, and a willingness to share their culture and traditions.

The Indonesian locals became his guides and friends, eager to showcase the beauty and uniqueness of their country. They shared stories of their ancestors, their customs, and their way of life, giving him a deeper understanding of the profound connections between the Indonesian people and their rich history. The locals' warmth and generosity became a source of inspiration for Hindolo, reinforcing his belief in the inherent goodness of humanity.

From Indonesia, Hindolo ventured into China, where he found himself captivated by the timeless wonders that bore witness to a civilization steeped in rich history, profound traditions, and remarkable ingenuity. The Great Wall stood as a magnificent architectural marvel, an enduring symbol of China's strength and ambition. Hindolo marveled at its grandeur, tracing his gaze along its serpentine path that disappeared into the distant mountains. The sheer magnitude of this colossal structure left him humbled, evoking a profound sense of wonder at the remarkable feat of engineering and construction achieved by the ancient Chinese civilization.

As Hindolo immersed himself in the captivating landscapes surrounding the Great Wall, he also connected with the people who called this land home. The Chinese people he encountered radiated a palpable sense of pride in their rich cultural heritage. They shared stories passed down through generations, tales that spoke of the indomitable spirit of the Chinese people and their unwavering commitment to preserving their cultural identity.

Through his interactions, Hindolo witnessed the deep-rooted traditions and values that permeated Chinese society. The people he met shared stories, rituals, and customs that carried the weight of history. Hindolo marveled at the meticulous rituals of tea ceremonies, the harmonious practice of martial arts, and the mesmerizing performances of traditional Chinese opera. Each aspect of Chinese culture seemed to exude a profound sense of reverence, demonstrating the depth of respect and gratitude the Chinese held for their past.

China became a captivating chapter in Hindolo's autobiography. This chapter of his life celebrated the resilience of the human spirit and the enduring legacy of a civilization deeply rooted in tradition and innovation.

With its complex history and indelible impact on Hindolo's soul, South Africa beckoned him once again. As he set foot on its soil, he found himself immersed in a land that bore the scars of apartheid, a painful reminder of a dark chapter in human history. Yet, alongside the remnants of segregation, he also witnessed the triumphs of unity and resilience—the very essence that defined the spirit of the nation.

Walking through the vibrant streets of Soweto, Hindolo could feel the weight of the struggle and the undeniable hope that pervaded the air. He listened to the stories of those who had fought relentlessly for freedom and equality, their voices carrying the strength of collective action. Hindolo witnessed the devastating consequences of injustice and inequality through his work with legal aid clinics, where he saw firsthand the faces of those he had helped—their tears mingling with gratitude and resilience etched in his memory.

However, it wasn't just the grand historical narratives that shaped Hindolo's understanding of South Africa. The everyday lives of its people—their struggles and triumphs—truly illuminated the nation's essence. He recalled an encounter in Guinea, where he met an elderly woman whose weathered hands told a tale of a life marked by resilience and unwavering determination. Her strength and perseverance became an embodiment of the human spirit, teaching him invaluable lessons about the power of resilience.

Hindolo's journey took him to Uganda, where he confronted the stark reality of poverty that permeated the lives of many. The contrast between the opulence enjoyed by a privileged few and the destitution faced by the majority served as a poignant reminder of the immense challenges that needed to be addressed. Yet, amidst the depths of poverty, he witnessed the unwavering generosity of the Ugandan people—their ability to come together and share whatever little they possessed. It was a powerful lesson in humility and the power of community.

Throughout his travels, Hindolo's encounters with different cultures, landscapes, and stories shaped his understanding of the world and deepened his commitment to justice, equality, and the uplifting of marginalized communities. His journey became an autobiography, a testament to the transformative power of human connections, empathy, and resilience.

As the sun reached its zenith, casting a warm, golden glow over the city, Hindolo took a deep breath, ready to embrace the next phase of his life's journey. The pages of his autobiography turned, filled with anticipation, the promise of a brighter future, and an unwavering belief in the possibility of change. Hindolo carried with him the wisdom gained from his experiences. The stories etched in his heart and a burning determination to make a lasting impact on his beloved Sierra Leone and the world beyond.

Hindolo's work had taken him to the turbulent land of Mexico. Attending a youth conference in Ixtapa, he landed in the beautiful land. Ixtapa ZTAPA is a Pacific Coast beach resort in the Mexican state of Guerrero. The coastline is bordered by curving El Palmar Beach has high-rise hotels, bars, and restaurants. As modernized, fishing boats leave from Ixtapa Marina, adjacent to a golf course. The granite rocks of Los Morros de Potosí are a popular dive site, while the Delfiniti Dolphinarium offers the chance to swim with dolphins. The laid-back resort city of Zihuatanejo lies just southeast.

Hindolo finds himself fascinated by a place resonating with an almost mystical vibe and in a pristine, privileged position, Ixtapa.

He met his old pal, Alison, whom he first—completed in Sierra Leone. Alison is a Los Angeles-based real financial impact crusader and loves Mexico. She explained to him that "the state of Guerrero is part of the Costa Grande region and is also one of the three cities named the "Triangle of the Sun" alongside Acapulco and Taxco de Alarcon. Given that Zihuatanejo was traditionally a quaint fishing village, the charm of the twinkling lights of the neighborhoods in the surrounding hills possesses an authentic vision of a simpler time and place. Swimming in the bay after sundown offers one of the most charming views worldwide."

Quite knowledgeable in its history, he won't say with pride that "the hills surrounding Zihuatanejo provide an almost amphitheater-like view in which the viewer is transported to a fairy village full of magical possibilities. Strolling the gorgeous new promenade parallel to the sea, holiday revelers are afforded incredible vistas all along the pathway that leads from Principal Beach to the quieter and more serene Madera Beach, where a slew of excellent outdoor restaurants will greet travelers and offer up exquisite seafood dishes and cold beverages to refresh and entice."

"Fascinating! Can you show me around? " Hindolo said.

"if we have time…" she responded.

"Ixtapa, the more developed of the two towns, offers many amenities that the quaint Zihuatanejo might not have. In Ixtapa, you'll likely find European delicacies and higher-level gastronomic delights. Ixtapa is where you'll find world-class golf courses, an excellent marina, and many incredible restaurants. Shopping in Ixtapa is also of a higher caliber than the more down-to-earth Zihuatanejo. With shopping centers, tattoo shops, burgeoning excursions & bike tours, Ixtapa provides a more contemporary experience for those not interested in "roughing it," she said.

Hindolo's mind raced back to Southeast Asia. He visited Indonesia to attend a conference on global democracy full of fascinating activists once. The story of the struggle for democracy worldwide can be thrilling and sometimes scary. Life-threatening when they question governance patterns or violations of human rights. A friend calls activists "tree huggers " and "insane" for advocating for pure everything - democracy, human rights, and love. The friend would say there is totally nothing. Indonesia is located off the coast of mainland Southeast Asia in the Pacific and Indian Oceans. An archipelago lies across the equator and spans a distance equivalent to one-eighth of Earth's circumference. Indonesia islands are grouped into the Greater Sunda Island of Sumatra (Sumatera), the southern extent of Borneo (Kakimantan), Java (Jawa), and Celebes (Sulawesi); the Lesser Sunda Islands (Nusa Tenggara) of Bali and a chain of islands that runs eastward through Timor; the Moluccas (Maluku) between Celebes and the island of New Guinea; and the western extent of New Guinea (generally known as Papua).

He landed in the capital, Jakarta, near Java's northwestern coast. In the early 21st century, Indonesia was the most populous country in Southeast Asia and the fourth most populated in the world. Hindolo was thrilled to have come here with just his passport and a letter that he was a VIP of the government.

His love for history and places pushed him to ask more questions and read much about Indonesia. Indonesia was formerly known as the Dutch East Indies (or Netherlands East Indies). The name lasted for a while and got changed to its present only of independence. A German geographer used Dutch East Indies as early as 1884; it is thought to derive from the Greek *indos*, meaning "India," and *nesos*, meaning "island." After a period of occupation by the Japanese (1942–45) during World War II, Indonesia declared independence from the Netherlands in 1945. Its struggle for freedom, however, continued until 1949, when the Dutch officially recognized the Indonesian state and sovereignty. It was not until the United Nations (UN) acknowledged the western segment of New Guinea as part of Indonesia in 1969 that the country took on its present form. The former Portuguese territory of East Timor (Timor-Leste) was incorporated into Indonesia in 1976. Following an UN- organized referendum in 1999, East Timor declared its independence and became fully state in 2002.

Interestingly, the Indonesian archipelago represents one of the most unusual areas in the world: it encompasses a significant juncture of Earth's tectonic plates, spans two faunal realms, and has for millennia served as a nexus of the peoples and culture of Oceania and mainland Asia. These factors have created a highly diverse environment and society that sometimes seem united only by susceptibility to seismic and volcanic activity, proximity to the sea, and a moist, tropical climate.

Many nationals believe a centralized government and a common language have united Indonesia. Furthermore, in keeping with its role as an economic and cultural crossroads, the country is active in numerous international trade and regional security groups.

As Hindolo continued his journey of reminiscence, his thoughts turned to significant historical moments that had profoundly impacted the world. One such moment was the creation and collapse of the Berlin Wall, which separated East Germany from West Germany during the height of the Cold War.

The Berlin Wall came into existence on the night of August 12–13, 1961, as a response to the mass exodus of East Germans to the West. About 2.5 million East Germans fled to West Germany between 1949 and 1961, including skilled workers, professionals, and intellectuals. This significant loss posed a threat to the economic viability of the East German state, leading to the construction of the barrier known as the Berlin Wall.

Initially built with barbed wire and cinder blocks, the Berlin Wall eventually evolved into a series of concrete walls, some as high as 15 feet, topped with barbed wire, and heavily guarded with watchtowers, gun emplacements, and mines. The wall stretched 28 miles through Berlin, dividing the city into two parts, and extended an additional 75 miles around West Berlin, separating it from the rest of East Germany.

The Berlin Wall became a poignant symbol of the division between East and West Germany and the broader divide between Eastern and Western Europe. It embodied the tensions of the Cold War era, where families were torn apart, and countless lives were affected by its presence.

During its existence, about 5,000 East Germans managed to cross the Berlin Wall, risking their lives to reach safety in West Berlin. However, many were not as fortunate, as East German authorities captured 5,000 others in their attempt, and tragically, 191 individuals lost their lives while attempting to cross the wall.

The Berlin Wall stood as a stark reminder of the struggles those living under oppressive regimes faced, the lengths people would go to seek freedom, and the courage required to defy the barriers that sought to confine them.

As Hindolo's journey continued, he couldn't help but reflect on the resilience and determination of those who lived through this tumultuous period of history. The echoes of the past reverberated in his mind, reminding him of the significance of standing up for justice and equality, even in the face of insurmountable odds.

The Berlin Wall's rise and eventual fall served as a testament to the power of collective action and the unwavering human spirit.

It was a pivotal moment in history that would forever be etched in the annals of time, inspiring generations to come.

As he ventured through the landscapes of his memories, Hindolo realized that history had shaped him, leaving an indelible mark on his beliefs and guiding him in his commitment to serving humanity. The stories of those who crossed the Berlin Wall and those who couldn't echo in his heart, reinforcing his determination to continue his fight for justice and equality in his beloved Sierra Leone and the broader global context.

As Hindolo delved deeper into the memories of Sierra Leone, he couldn't help but be reminded of its rich history as part of British West Africa. The assortment of territories in western Africa, including Sierra Leone, the Gambia, Nigeria, and the Gold Coast, had been administered by Great Britain during the colonial period.

Sierra Leone's colonization had a unique origin, with formerly enslaved people arriving from England in 1787, followed by groups from Nova Scotia in 1792 and Jamaica in 1800. They were sponsored and governed by the private Sierra Leone Company until 1808, when Britain made Sierra Leone a crown colony. The establishment of the colony of Bathurst at the mouth of the Gambia River in 1816 was another significant milestone in British West Africa. Both colonies served as bases for the British effort to block the slave trade along the coast, aligning with the broader abolitionist movement.

As the British rule spread to the interior of Sierra Leone and the Gambia, both regions became protectorates governed by indigenous rulers. The British policy of indirect rule, formulated most clearly by Frederick J.D. Lugard in Nigeria, became the model for all of British West Africa. This system allowed local administration and jurisdiction to depend on traditional rulers and institutions, facilitating governance while preserving existing social structures.

Parts of the Gold Coast, present-day Ghana, were also acquired by Britain at different times. The Gold Coast crown colony was established in 1874 in Fante and Ga lands near the British coastal trading forts. The Asante empire to the north was conquered and made a protectorate in 1900–01. British West Africa saw significant changes throughout the years, including the division of the former German colonies of Togoland and Kamerun between Britain and France as League of Nations mandates after World War I.

However, as Hindolo traversed through his memories, he became increasingly aware of the complexities of British rule in the region. The policy of indirect rule, while preserving traditional systems, also excluded Western-educated Africans from positions of power and suppressed social change. This disparity gave rise to nationalist movements for independence, led by Western-educated Africans seeking self-governance and representation.

As the winds of change swept through British West Africa, Ghana became the first to gain independence in 1957, followed by Nigeria in 1960, Sierra Leone in 1961, and The Gambia in 1965. The British Cameroons were divided between Nigeria and the Republic of Cameroon. The end of British West Africa marked a new chapter in the history of the region, as its people embarked on a journey towards self-determination and nation-building. Hindolo's own life journey was now interwoven with the collective struggle for a brighter future, where the echoes of the past fueled the determination to create a just and equitable society for all.

CHAPTER 10

MAN'S DESIRE FOR POWER

As Hindolo's journey through his memories continued, he found himself contemplating the intricate facets of human nature, particularly the unquenchable thirst for power. Throughout his life, he had encountered individuals who had once been close friends, only to succumb to the allure of power, forsaking the fundamental principles of humanity.

History is replete with instances where remarkable individuals have been consumed by their relentless pursuit of power, an insatiable drive to ascend to the summit through any means necessary, even if it entails toppling mountains and shattering moral boundaries to grasp the elusive concept of power. Political authority encompasses the capability to shape the conduct of people and prized assets, influencing a society's policies, functions, and ethos, often bolstered by military might.

However, in the context of African politics, particularly Sierra Leone, Hindolo frequently pondered whether the trajectory of his homeland could have taken a divergent course had individuals adhered to influencing societal policies, functions, and culture rather than indulging in self-enrichment at the expense of the collective.

As he progressed along his reminiscences, Hindolo immersed himself in the tales of those he had once held dear – individuals whose paths had veered away from the pursuit of justice and compassion, enticed instead by the allure of dominance and supremacy. These were comrades who had stood shoulder to shoulder with him in the battle for a more just world, allies who had ultimately chosen a contrasting route.

One vivid recollection etched itself into Hindolo's memory – that of a friend who had once been an ardent champion of human rights and societal fairness. Together, they had navigated the tumultuous seas of activism, locked arm in arm in the crusade for the oppressed and marginalized. Their shared aspiration for a more egalitarian society had kindled an unquenchable fire within them, cultivating a bond built on camaraderie and shared purpose.

Yet, the passage of time unfurled subtle shifts in his friend's demeanor, unsettling shifts that signified the creeping influence of ambition and the seductive pull of power. What was once a steadfast commitment to uplifting the downtrodden had morphed into an insatiable hunger for status and authority. Hindolo watched with a heavy heart as his friend gravitated towards those in positions of influence, courting favor and aligning with those in authority to secure personal gain.

Soon enough, Hindolo recognized that his friend's unbridled craving for power had led them astray, ensnaring them in a perilous labyrinth. In their quest for dominance, moral considerations seemed to wane in significance. The once-clear demarcation between right and wrong grew hazy, with the essence of humanity becoming a faint echo in the background, overshadowed by the pursuit of supremacy.

Hindolo comprehended that power was a complex entity capable of both constructive and destructive outcomes. While some harnessed its potential to drive positive change, others were vulnerable to its corrupting influence, their ideals corroded by the allure of authority. This transformation was palpable in his friend, once a fervent advocate, now seemingly intoxicated by the taste of power. Their actions took on a ruthless character, ready to embrace extreme measures in their ascent to dominion. Hindolo was reminded of the profound wisdom of the ancient Greek philosopher Aristotle, who mused, "Anybody can become angry – that is easy; but to be angry with the right person, to the right degree, at the right time, for the right purpose, and in the right way – that is not within everybody's power and is not easy."

In response, Hindolo felt compelled to engage with his friend, hoping to rekindle the spark of their shared ideals and steer them away from the treacherous path of unchecked power. It was a task imbued with challenges, yet he remained steadfast in his belief that the essence of humanity could prevail over the allure of dominance. As he continued to reflect on these experiences, Hindolo recognized the delicate balance that existed between wielding power responsibly and succumbing to its corrosive temptations, a balance that would forever shape the trajectory of individuals and nations alike.

As Hindolo's journey through his memories unfolded, one particular figure occupied his thoughts—Santigue, a young man with an ebony complexion who had been hailed as a fervent firebrand. Santigue was uniquely able to converse in various indigenous languages, making him a potential activist capable of connecting with diverse communities. Hindolo vividly recalled the moment Santigue had confided in him, stating, "I would rather perish from hunger than become entangled in politics. Politics is a realm of malevolence, reserved for the avaricious." These words resonated deeply with Hindolo's staunch commitment to speaking truth to power, forging an instant bond between the two.

Santigue's evolution over a decade spanned continents, as he secured an international activist scholarship to study in Europe. Upon his return, however, a disconcerting transformation had occurred. The once-principled Santigue had succumbed to the allure of leadership, but with a twist—his path now intertwined with political parties, the very entities he had denounced. Hindolo was taken aback and initially thought it was a jest until he witnessed Santigue's active participation in one of the country's major political factions. The disillusionment was profound; his friend, once an embodiment of steadfast values, had seemingly reversed his convictions.

The disheartening trend of betrayal and manipulation in the pursuit of personal ambitions further revealed itself. Hindolo grappled with the stark dichotomy between the past and present of those he once held in high regard. The erosion of their commitment to the ideals of humanity weighed heavily on him, a poignant reminder of the fragility of human nature. He yearned for a return to the shared values that had united them in the fight for justice and compassion.

One chilling incident involving Mohamed, a friend turned political rival, stood as a stark exemplification of the corrosive effects of power. Mohamed's insatiable desire for supremacy led him to plot against Jacob, a competitor, resulting in a heinous vehicular tampering that gravely injured Jacob and cost him a limb. Hindolo's shock was palpable, as Mohamed's descent into ruthlessness and manipulation stood in stark contrast to his college days as a compassionate advocate.

Hindolo was resolute in his belief that ambition, while commendable, could become a dangerous force when unchecked. He candidly conversed with Mohamed, stressing the potential pitfalls of excessive ambition. However, Mohamed's response, "the end justifies the means," showcased the dangerous extent to which power could warp one's moral compass. This Machiavellian perspective shocked Hindolo and emphasized the alarming shift in his friend's character.

The narrative took a turn towards introspection as Hindolo grappled with the nuances of ambition, insecurity, and their effects on human behavior. He shared his insights with Mohamed, urging him to temper his ambitions to avoid a path of reckless overreach. However, Mohamed, who asserted his willingness to employ any means necessary to achieve power, met Hindolo's efforts with defiance and dismissal.

In a world where power often became a tool for manipulation, Hindolo's belief in the inherent goodness of humanity was tested. His experiences with friends like Santigue and Mohamed left him questioning how individuals could stray so far from their core values in the pursuit of dominance. Yet, amidst the disillusionment, Hindolo found solace in individuals like Anthony, an unwavering activist who fearlessly spoke truth to power without compromise. Anthony's steadfastness provided a glimmer of hope, challenging the prevailing notion that ethical leaders were a rarity.

The chapter underscored the profound impact of ethical leadership and the preservation of humanity's creed. It served as a rallying cry to uphold the principles of compassion, empathy, and justice, even in the face of overwhelming power dynamics. Hindolo recognized that the allure of power had the potential to compromise even the noblest intentions, emphasizing the urgency of ethical leadership and its role in shaping a just society.

As Hindolo's reminiscences came to a close, he acknowledged the delicate equilibrium required in the pursuit of power. He resolved to continue his advocacy for a better world, determined to safeguard the values that had once united him with friends like Santigue and Mohamed. Armed with the lessons of his encounters, Hindolo moved forward with a renewed sense of purpose, striving to create a legacy that would transcend the allure of power and endure as a testament to the enduring strength of humanity's creed. Time, he reflected, might crumble the edifices of power, but the ideals of compassion and empathy would stand as a testament to the resilience of the human spirit.

CHAPTER 11

TRAVERSING THE UNKNOWN

Countless challenges, triumphs, and introspective moments had marked the journey of Hindolo. From rage to almost riches, from insignificance to a prominent figure on whom the welfare and sometimes the safety of others rest. The country's future is in the hands of him and his contemporaries, who are becoming the crème de la crème of the Sierra Leonean society. History is replete of a two sides path in this quagmire: either succeed or betray it. There will be no excuses for the lack of the opportunity to play their part in shaping the battered history of a land ostensibly full of honey and potential but arid in reality. As he stood at the unknown threshold, a new chapter unfolded before him—one that would test the core of his convictions and lead him into uncharted territory. This new path delved into the uncertainties that lay ahead, the internal struggles of a hero torn between different paths, and the unwavering resilience of the human spirit.

In the midst of swirling doubts and conflicting desires, Hindolo found himself at a crossroads. The allure of power and politics tugged at his ambitions, whispering promises of influence and change on a grand scale. Yet, his commitment to humanity's creed remained steadfast, a beacon illuminating his moral compass and guiding his decisions.

The uncertainty that shrouded his path was palpable. Each step he took seemed to lead him further into a realm of unknown possibilities, where the fog of doubt obscured the right choice. The hero within him was left in a state of perpetual contemplation, wrestling with the weight of his decisions and the consequences they might bring.

As he journeyed on, Hindolo's inner turmoil was laid bare. His story delved into his moments of indecision. His sleepless nights were spent pondering the ramifications of his choices and the conversations he held with mentors and confidantes. A complex interplay of hope and fear, ambition and ethics churned within his heart.

One recurring theme was the constant temptation he faced. The corridors of power were riddled with pitfalls, and Hindolo encountered numerous opportunities to compromise his principles for personal gain. Yet, time and again, he resisted the allure of corruption. His inner battles, showcasing his strength of character, enabled him to walk away from the edge of moral compromise.

Amidst the uncertainty, Hindolo found solace in his role as an educator. He accepted a position at the university, embracing the opportunity to impart knowledge, shape young minds, and foster a sense of critical thinking among the next generation. His interactions with eager students became a source of inspiration, reminding him of the profound impact that a single individual could have on shaping the future.

Standing before his students, he shared stories of his journey, the places he had seen, the people he had met, and the lessons he had learned. Through his words, he encouraged them to question, challenge, and remain steadfast in their pursuit of a just and equitable world. The classroom became a space where his ideals found fertile ground, where the seeds of change were sown in the minds of those who would carry the torch forward.

Yet, the call of politics remained ever-present. The path forward chronicled Hindolo's gradual immersion into the realm of governance as he navigated the intricacies of political campaigns, forged alliances, and engaged in passionate debates. The hero's evolution from an advocate to a potential lawmaker was uncovered with nuance, capturing the complexities of the political landscape and the personal sacrifices it demanded.

Throughout this tumultuous journey, one thing remained clear: humanity's creed was the North Star that guided Hindolo's every move. The values of compassion, justice, and empathy continued to shape his decisions, even as he ventured into unfamiliar and treacherous waters. His unwavering commitment to these principles served as a powerful reminder that the pursuit of power must always be tempered by a responsibility to uplift and serve the greater good.

As his story slowly came to a close, he traversed the unknown with grace and determination. Hindolo's path had been fraught with challenges, and he had been tested in ways he could never have anticipated. Yet, his journey was a testament to the resilience of the human spirit, the transformative power of ethical leadership, and the enduring impact of one individual's unwavering commitment to humanity's creed.

The story concluded with a sense of anticipation, hinting at the possibilities that lay ahead. While the road ahead remained uncertain, Hindolo's resolve remained unshaken. As he stood on the precipice of a new phase in his journey, the hero within him was resolute, ready to continue the pursuit of a more just and equitable world. The final pages of his incredible story were filled with a sense of hope and promise, a reminder that even in the face of uncertainty, the human spirit had the power to transcend, transform, and transverse the unknown.